D1312517

Weed Control Methods for River Basin Management

Author

Edward O. Gangstad

Aquatic Plant Control Botanist
Army Corps of Engineers
Washington, D.C.

Published by

CRC PRESS, Inc.
2255 Palm Beach Lakes Blvd.·West Palm Beach, Florida 33409

Library of Congress Cataloging in Publication Data

Gangstad, Edward O.
   Weed control methods for river basin management.

   Bibliography: p.
   1. Aquatic weeds – Control. 2. Watershed management. I. Title.
SB614.G36   628.9   77-24977
ISBN 0-8493-5328-9

International Standard Book Number 0-8493-5328-9

Library of Congress Card Number 77-24977
Printed in the United States

PREFACE

Aquatic plants continue to create problems associated with navigation, flood control, agriculture, irrigation and drainage, values of lands, conservation of wildlife and fisheries, and water resource supply. While much research is being done to find more effective and economic control measures, there is now a great need to apply known facts to achieve a measure of control by the means available. It is the purpose of this volume to provide a scientifically documented treatise of the known facts as they apply to the control of aquatic weeds in river basins and their allied waterways, with particular emphasis on alligator weed and water hyacinth.

Edward O. Gangstad

THE AUTHOR

Dr. E. O. Gangstad, Aquatic Plant Control Botanist, transferred to the Office of the Chief of Engineers in October of 1966 from the Texas Research Foundation, Dallas, where he was employed as Principal Agronomist of the Hoblitzelle Agricultural Laboratory. Dr. Gangstad was graduated from the University of Wisconsin with an M.A. in Biochemistry (1947) and from Rutgers University with a Ph.D. in Agronomy (1950) and minors in Plant Pathology and Plant Physiology.

During the past 25 years he has established an outstanding career in the environmental and biological fields and has over 80 scientific publications. He is a contributor to the *Journal of Biological Chemistry, Agronomy Journal, Crop Science, Journal of Heredity, Weed Science, Crops and Soils, Tropical Agriculture, Turrialba, Soil Science Society of Florida, Hoblitzelle Agricultural Bulletin, Wrightia, Hyacinth Control Journal, Environmental Quality Journal,* and the Journals of the American Society of Civil Engineers.

Dr. Gangstad's experience includes research and field studies on the physiology and pathology of plants at the Universities of Wisconsin and New Jersey; research and development studies on strategic long-vegetable fibers and related programs in Florida with the United States Department of Agriculture, Office of Naval Research, and Department of Defense; and agrobusiness studies to develop renewable natural resources of the State of Texas with the Texas Research Foundation and management of natural resources for the Department of the Army.

Dr. Gangstad is listed in *American Men of Science, Leaders in Science,* and *American Men of Achievement.*

IN MEMORIAM

This book is dedicated to Charles M. Schwartz, an employee of the Army Corps of Engineers, who died of cancer May 3, 1977 at his home in Alexandria, Virginia

TABLE OF CONTENTS

Part I
Environmental Survey and Assessment

Chapter 1

INTRODUCTION

Aquatic plants are an essential feature of the aquatic ecosystem. Benefits of aquatic plants in the ecosystem include: (1) production of oxygen through photosynthesis, (2) cover for young fish to escape predation, (3) sites for attachment by a number of organisms, (4) protection against wave erosion, (5) food for wildlife and waterfowl, and (6) removal of nutrients from the water.

When aquatic plants are in an overabundance in a body of water, they seriously interfere with water use. Problems associated with dense plant infestations are, (1) reduction of flow in irrigation and drainage canals, (2) suppression of growth of desirable plants, (3) clogging of intake pipes and passageways, (4) interference with fish harvesting, (5) creation of favorable habitats for growth and development of disease-carrying organisms, and (6) socioeconomic limits to area development.

The task of this volume has been to assemble and correlate various bits of information in the statistical sense, to present to the reader an integrated whole. The authors have drawn freely from the published record of research and development of the past three decades, and an effort has been made to include unpublished information with due credit to the original author. Laboratory and field research studies are largely limited to alligator weed and water hyacinth.

BASIC CONCEPTS

The science of ecology may be simply defined as the study of the interrelationships among biological organisms and their influence on, or their reaction to, the environment. Ecologic investigations may take the form of studying the response of a single species to the environment (autecology), or the interactions of a community of species (synecology). More recently, the study of ecology has been approached on an ecosystem basis which involves the capture and storage of radiant energy by green plants and its subsequent transformation and reduction by the biological and environmental systems. It is important to remember that the ecosystem involves not only the biological organisms, but also all aspects of the physical environment in which those organisms live.

The ecosystem approach to ecologic thought is a modern concept that has evolved in response to needs that were not met by the more classical concepts. Despite the interpretive excellence of an ecosystem analysis, it will not be used in this report because the quantitative data needed for such an analysis are not available. It is understood, however, that an effect on one element of an ecosystem cannot be isolated, and that each effect will be an influencing factor on one or several other aspects of the ecosystem, including the domain of man. Cultural changes due to water resource development on biota, soils, and climate are also discussed. However, synthesis of various parts into a fully interpretable whole is limited, and there are certain questions of importance about which responsible judgments cannot be made without further research.

METHOD OF STUDY

The discussion of possible ecologic consequences of water resource development are based upon evidence in the published literature, prior experience of the ecology of temperate and tropical climates, and general experience in the operation and maintenance of established water resource programs. Observations for this survey of different water resource development areas were made by first viewing the vegetation from a helicopter over-flight. More detailed information was obtained in the course of being transported by landcraft on the ground. Several hours were spent at each location to assess the successional patterns and to estimate possible effects on the relative density and compostion of the vegetation. Aerial and ground surveys were concentrated in large areas where ecological consequences of water resource projects could be most readily defined. The conclusions are limited to the observations that were possible within the time frame available.

POTENTIAL PROBLEMS

Large-scale modification of vegetation for water resource development is thought to cause changes in climate. The theory is that as the forest or grassland is converted to cropland, the evapotranspiration surface is reduced so that precipitation will be reduced. For this to happen, however, major changes in the vegetative cover would have to occur, much larger than is normally involved in a water resource project.

Another potential problem is the increased laterization of the soil. This refers to an indurated concreationary soil deposit high in iron or aluminum oxide. True laterite hardens irreversibly, and it will not soften upon wetting, making the soil unsatisfactory for cropping. In the tropics this tends to follow when these soils are cultivated, as may be the case under a water resource development program.

Soil erosion may also be a problem associated with new water resource developments. Soil type, topography, vegetative cover, and rainfall intensities are important factors related to soil erosion. In general, erosion will be greatest on porous sandy soils with steep slopes and reduced vegetative cover. In a tropical climate, it is also known that soil nutrients are readily leached, particularly where the soil is cultivated. This adds to the erosion problem. In temperate climates, these problems are much less severe.

RESOURCE INFORMATION

In the development of the following text, the author wishes to acknowledge sources of information cited.

Chapter 2 — House, W. B., Goodson, L. H., Gadberry, H. M., and Dockston, K. W., Assessment of Ecological Effects of Extensive or Repeated Use of Herbicides, Mimeograph Report, Department of the Army, Washington, D.C., 1967.

Chapter 3 — Divisions of River Basin Activities, Mimeograph Report, U.S. Department of the Interior, Washington, D.C., 1973.

Chapter 5 — Consequences of Building Dams, U.S. Army Corps of Engineers, Washington, D.C., 1973.

Chapter 6 — Nelson, M. L., Seaman, D. E., and Gangstad, E. O., Potential Growth of Aquatic Plants of the Lower Mekong River Basin, Laos-Thailand, U.S. Army Corps of Engineers, Portland, Ore., 1970.

Chapter 7 — Gangstad, E. O., Seaman, D. E., and Nelsen, M. L., Potential Growth of Aquatic Plants in the Republic of the Philippines and Projected Methods of Control, Office of the Chief of Engineers, Washington, D.C., 1971.

Chapter 8 — Gangstad, E. O., Seaman, D. E., and Nelson, M. L., The Potential Growth of Obnoxious Plants and Practical Systems of Control in the Republic of India, Office of the Chief of Engineers, Washington, D.C., 1971.

Chapter 9 — Turburg, F. W., Forch, J. A., Solmosy, S. L., and Hays, S. A., Control of Alligatorweed and Other Aquatic Plants, Mimeograph Report, University of Southwestern Louisiana, Lafayette, 1967.

Chapter 14 — Blackburn, R. D. and Durden, W. C., Integration of Biological and Chemical Control of Alligator Weed, Cooperative Study of the U.S. Army Corps of Engineers and the U.S. Department of Agriculture, Fort Lauderdale, Fla., 1973.

Chapter 16 — Water Hyacinth Obstructions in the Waters of the South Atlantic States, House Document No. 37, U.S. Government Printing Office, Washington, D.C., 1957.

Chapter 19 — Expanded Project for Aquatic Plant Control, House Document No. 251, U.S. Government Printing Office, Washington, D.C., 1965.

Chapter 21 — Barry, J. R. and Forest, J. S., Time Course Studies, University of Southwestern Louisiana, Lafayette, 1970; Leiger, C. F., Punta Gorda Program, U.S. Army Corps of Engineers, Jacksonville, Fla., 1970.

Chapter 2
ECOLOGY OF AQUATIC WEED CONTROL*

INTRODUCTION

An ecosystem involves the capture and accumulation of energy and matter, and their circulation and transformation through the medium of biotic (living) things and their activities. Photosynthesis, herbivore action, predation, decomposition, parasitism, and symbiotic activities are among the principal biological processes responsible for the transport and storage of materials and energy, and the interactions of the organisms engaged in these activities provide the pathways of distribution. (A glossary of ecological terms for description of these phenomena is given in the Appendix.)

In the abiotic (nonliving) part of the ecosystem, circulation of energy and matter is completed by such physical processes as evaporation and precipitation, erosion, and deposition. The ecologist, then, is primarily concerned with the quantities of matter and energy that pass through a given ecosystem and with the rates at which they do so. Of almost equal importance, however, are the kinds of organisms that are present in any particular ecosystem and the roles that they occupy in its structure and organization.

Ecosystems have a complex of regulatory mechanisms, which, in limiting the numbers of organisms present and in influencing their physiology and behavior, control the quantities and rates of movement of both matter and energy. Processes of growth and reproduction, agencies of mortality (physical as well as biological), patterns of immigration and emigration, and habits of adaptive significance are among the more important groups of regulatory mechanisms.[2]

DEFINITION OF AN ECOSYSTEM

The assemblage of plants and animals in an ecosystem usually consists of numerous species, each represented by a population of individual organisms. However, each population can be regarded as an entity in itself, interacting with its environment (which may include other organisms as well as physical features of the habitat) to form a lower level of organization that likewise involves the distribution of matter and energy. Each individual animal or plant, together with its particular microenvironment, constitutes a still lower level of organization.

All ranks of ecosystems are open systems, not closed ones. Energy and matter continually escape from them in the course of the processes of life, and they must be replaced if the system is to continue to function. The pathways of loss and replacement of matter and energy frequently connect one ecosystem with another, and therefore it is often difficult to determine the limits of a given ecosystem. At whatever level we may wish to consider, the ecosystem stands as a basic unit of ecology.

The whole complex of the plants and animals forming a community, together with all the interacting physical factors of the environment, are a single unit. This takes into account all the living creatures in the community; the fungi, bacteria, and nematode worms living in the soil, the mosses, caterpillars, and birds in the trees, and all the factors of the environment, including the composition of the soil atmosphere and soil solutions, wind, length of day, relative humidity, atmospheric pollution etc., have some effect on the balance of the whole. The final aim of ecologists is the complete understanding of ecosystems, an ideal one cannot really hope to attain. It is nevertheless an ideal well worth pursuing, and progress has been made toward it. The first stage is to simplify the system by studying different aspects of the ecosystem independently.

A forest, a lake, a savannah, a marsh, a desert, and a prairie are all examples of ecosystems at the community level. Although each may be a different composition of environmental and biological factors, the basic structure, within which matter and energy transfers are made, is very much the same.[3,9]

All functions, and indeed all life, within an ecosystem depend upon the utilization of an external source of energy, solar radiation. A portion of this incident energy is transformed by the process of photosynthesis into chemical energy in the structure of living organisms. In the language of community economics, autotrophic

*From House, W. B., Goodson, L. H., Gadberry, H. H., and Dockton, K. W., Assessment of Ecological Effects of Extensive or Repeated Use of Herbicides," Department of the Army, Mimeograph Report, 1967.

plants are producer organisms, employing the energy obtained by photosynthesis to synthesize complex organic substances from simple inorganic substances. Although plants again release a portion of this potential energy in catabolic processes, a great surplus of organic substance is accumulated. Animals and hetero-trophic plants, as consumer organisms, feed upon this surplus of potential energy, oxidizing a considerable portion of the consumed substance to release kinetic energy for metabolism, but transforming the remainder into the complex chemical substances of their own bodies. Following death, every organism is a potential source of energy for saprophagous organisms (feeding directly on dead tissues), which again act as energy sources for successive categories of consumers. Heterotrophic bacteria and fungi, representing the most important saprophagous consumption of energy, may be conveniently differentiated from animal consumers as specialized decomposers of organic substance. It has been suggested that certain of these bacteria be further differentiated as transformers of organic and inorganic compounds. The combined action of animal consumers and bacteria decomposers dissipates the potential energy contained in organic substances while transforming them to the inorganic state. Figure 2.1 illustrates the relationship among consumer, producer, and scavenger zones.

BASELINE STUDIES

A thorough study of the ecosystem should be made as nearly in its unaltered condition as is possible. This would involve population studies of all the species of plants and animals. Measurements of the total biomass and biotic potential should be made. In spite of the logarithmic mode of current scientific advances, no reasonably comprehensive and balanced scheme of the ecology of a single "natural" ecosystem exists, nor is one likely to be produced in the near future. Although our purpose is to stress the fundamental unity of ecosystem physiology, it is impossible with the means at hand to describe the many interrelationships which exist in a natural ecosystem as well as their numerous and diverse interconnections.[3]

The quantitative determination of the biotic potential of any ecosystem appears to be beyond our current scientific capability, since the biomass, though measurable, is not a completely satisfactory estimation of net primary productivity. In addition Ovington[9] stated that it is impossible to generalize on long-term trends in the nutrient budget. This status of the nutritional "balance" persists, because the many diverse processes involved in chemical element circulation are interdependent functions of (1) the character of the ecosystem, (2) the particular chemical nutrient concerned, and (3) the influence of man.

ECOLOGICAL EFFECTS

Many thousands of different kinds of plants cover practically every available space on the surface of the earth. These plants, from the simplest unicellular algae to the largest sequoia, are able to take water and carbon dioxide, utilize the radiant energy from the sun, and in the presence of chlorophyll, convert them into simple sugars. Some of the simple sugars are compounded into starches, celluloses, and other carbohydrates. Others are changed into proteins, fats and oils, various pigments, resins, gums, pectins, and other substances by the addition of minerals (from the water medium or from the soil) and the utilization of the energy tied into the total process by photosynthesis. It is estimated that the plants of all the world's oceans fix 15.5×10^{10} tons of carbon per year; whereas land plants fix only 1.9×10^{10} tons of carbon per year into complex food compounds.[7]

One of the generalities widely accepted in ecology is that no two closely related species can occupy the same niche. One of them is better able to take its requirements from a particular environment than the other. If, however, they are found growing and thriving very closely together, then their requirements are sufficiently different, such that adverse competition does not occur. Instead, the relationship may be one of commensalism, mutualism, or neutralism. Some plants are able to grow and thrive in a wide variety of habitats, while others are restricted to very narrow niches in the biosphere. Basically, however, the vegetation is a direct result of the environment in which it grows. It is controlled by such abiotic factors as moisture, temperature, intensity and quality of light, humidity, soil, soil minerals, altitude, and wind, and such biotic factors as other plants, animals, parasites, etc.

In the normal course of events, plants die because they have completed their life cycle, have been unable to compete successfully with other

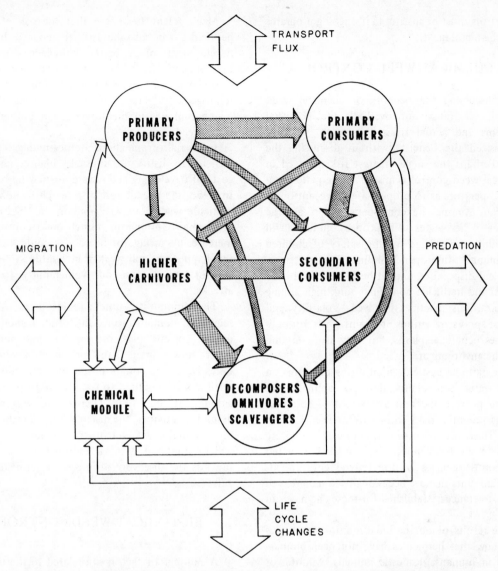

FIGURE 2.1. Consumer, producer, and scavenger zones. Solid arrows indicate intra-web flux by predation and feeding; open arrows indicate other transport and transfer within the food web or between food webs of different regions or zones. (From Gillet, J. W., Hill, J., Jarninen, A. W., Schoor, W. P., Ecological Research Series, EPA-660/3-74-024, 1974.)

plants, have been destroyed by disease, fire, or frost, or have been eaten by animals. Fortunately, the space which they occupied is then taken over by some other plant; whether that space is small or large, whether it requires one plant or many, it (the ecological niche) soon becomes filled.[6]

HYDROLOGIC CYCLE

The climate, microclimate, and weather are important factors affecting river basin assessments and development of aquatic plants related to these conditions. The ocean basins form a giant evapora- tion surface from which water is evaporated to form clouds and returned to the earth as rain. Solar energy is the dynamic power of this system.[4,10]

AQUATIC PLANTS

The biotic factors of the ecosystem provide the environment for growth of aquatic plants. Specific management of excessive growth of aquatic plants is frequently necessary under particular conditions that stimulate their growth. As a result, some form of controls must be instituted if ongoing programs are to be maintained.[10]

CHEMICAL WEED CONTROL

In discussing the benefits of chemical weed control, we must not overlook the potential problems and actual risks. Too often, we have emphasized the benefits without discussing the risks. Perhaps the single greatest risk involved in chemical weed control is misuse. Inadequate training of operators as to proper safety precautions is common. Among the commonest errors are the transfer of herbicides to improper containers and the failure to provide for the safety of children and animals. Improper handling, storage, transport, and disposal of unused herbicides, herbicide wastes, and herbicide containers rank high among the hazardous practices involved. Volatility and drift of sprays are serious problems. Drift not only damages beneficial plants, but may also contaminate the environment.[5,8]

The national pesticide monitoring programs on food, water, soil, plants, domestic animals, and wildlife provide the best assessment of actual as well as potential risks from current use of herbicides. These results show that herbicides are rarely found in the food monitoring studies conducted by various agencies of the federal government. Trace amounts that are occasionally found are well below tolerances established for these chemicals in food.[12]

The results of our national monitoring studies also show that herbicides have not been biomagnified in humans, domestic animals, wildlife, or other objects or organisms in the environment. There is no evidence that agricultural uses of herbicides cause an accumulation of residues in any of these organisms. Herbicides used at agricultural rates on croplands dissipate rapidly. There is little evidence from our monitoring programs of any accumulation or buildup of herbicidal residues in soils.[12]

It is genetically possible, and there is some academic concern that weeds may become resistant to herbicides, but most scientists believe that this is not a significant problem. There is greater concern over the increase in weed populations that are resistant to control by herbicides. Although these ecological shifts may involve some risks, they will not become serious if cultural, mechanical, ecological, biological, and chemical methods are combined in an integrated systems approach.

Many scientists believe that the risk of using herbicides can also be further reduced by the development of a better knowledge of their toxicological effects. More emphasis is being given to understanding the carcinogenic, mutagenic, and teratogenic effects of herbicides during the research and development phase and before their widespread use.[8]

Industrial organizations are beginning to find it difficult to justify the unusually high commercial costs of developing and registering new herbicides. In 1956, the overall cost of research to develop a pesticide was about $1.2 million. In 1972 it cost more than $4 million. If such costs continue to increase, a shortage of herbicides could occur. If this happens, it will weaken integrated systems of control and increase our vulnerability to weed attacks.[1]

The amount of time and resources used by scientists, administrators, and other personnel to defend current registered uses of herbicides and other unwarranted bans on uses has become excessive and counterproductive. If industrial scientists are required to spend a disproportionate amount of their time, talents, and resources in defensive activities, the discovery and development of new herbicides and other pesticidal chemicals will be inhibited. Many already believe that work towards the discovery of improved herbicides has suffered.[1,4,11,12]

BIOLOGICAL WEED CONTROL

A number of factors are related to the uncontrolled growth of aquatic macrophytes. The introduction of exotic plant species has resulted in rapid spread and colonization by many of these plants. In many cases these introduced species will dominate a body of water at the expense of native species. These exotic species generally are not a problem in their native habitat because natural enemies and other causes keep their growth under control. However, once these plants are transported to areas not containing these controls, the plants multiply rapidly. Herbicides are presently widely used to control growth of unwanted aquatic plants. Use of herbicides is generally only a temporary solution to the problem since many of these plants regrow quickly once the parent plant has been eliminated. Until better control methods are available, herbicides will continue to be used.

Biological control of aquatic plants appears to

have great potential, but has been largely unsuccessful because of the difficulty in finding organisms which can effectively reduce the growth rate of the aquatic plants. Use of the alligator weed flea beetle (*Agasicles hygrophila* Selman & Vogt) to control alligator weed (*Alternanthera philoxeroides* [Mart.] Griseb.) is one example of the use of an organism to control aquatic plant growth. However, control of alligator weed by this organism is limited to certain situations. An understanding of the organism as well as the aquatic plant is essential in order to gain maximum benefits from biological control of aquatic vegetation.[8,11]

The use of herbivorous fish to control aquatic vegetation has received considerable attention in recent years. Fish which feed partly or entirely on aquatic vegetation include Congo tilapia (*Tilapia melanopleura* Dumeril), Java tilapia (*Tilapia mossambica* Peters), Nile tilapia (*Tilapia nilotica* L.), *Tilapia zillii* Gervais, silver dollar (*Metynnis roosevelt* Eig.), *Mylossoma argenteum* E. Ahl, common carp (*Cyprinus carpio* L.), the Israeli strain of the common carp, and the white amur (*Ctenopharyngodon idella* Val.).

The University of Florida is placing more emphasis on the aquatic ecosystem than any other university or governmental agency in the United States. Florida's subtropical environment and abundance of water, plus its rapidly expanding population, provide excellent opportunities for study of both the natural aquatic ecosystem and its alteration by human population. Studies at the University are designed to gain a better understanding of the complex aquatic system while providing answers for control of aquatic plant problems with biological, chemical, and mechanical methods supplement with the utilization of unwanted plant growth.

Models which serve as the basis for these studies are described by Odum.[10] Energy flow models of this type are easily converted to mathematical equations for simulation. A simplified representation of energy flow pathways through the basic components of an aquatic ecosystem includes sunlight, water level, and oxygen as the sources of energy influencing plant growth. The supply of mineral nutrients affects the ability of plants to utilize these sources of energy. Some of the vegetation is consumed by animals; the rest of the vegetation as well as the animals are eventually broken down by decomposers into mineral nutrients which, together with nutrients from sources outside the system under investigation, help regulate plant growth.

The one specific control mechanism of interest in this study is the use of the white amur to feed on problem aquatic plants. A flow diagram showing the energy expenditures involved in the use of herbivorous fish may be devised. By using models of this type, it is possible to simulate trends which would occur after introduction of the fish to control aquatic plants. Energy expenditures involved in using the white amur include introducing and managing populations of this fish and native fish species to assure the maintenance of a proper balance in the system, harvesting and processing the fish, and the energy supplement provided by the harvested fish.[3] The Corps of Engineers is sponsoring a study of the white amur for control of aquatic vegetation at Lake Conway, Florida for the period 1976–1979.

Chapter 3

ENVIRONMENTAL IMPACTS FOR FISH AND WILDLIFE MANAGEMENT*

INTRODUCTION

A proper perspective is necessary for a valid concept of chemicals and their effects on the aquatic environment. Water alone, without chemicals, is a sterile environment unable to support most beneficial uses. Life can survive only after the addition of a specific combination of chemicals. Only then can selected organisms grow, reproduce, and sustain a population. A critical relationship in the chemical character of water exists with respect to both the flora and the fauna it will support. The diurnal and even temporary seasonal shifts or changes in chemical and/or physical characteristics of this environment create serious limiting factors to the populations sustained.

The development of modern biocides has brought about a new and very important change in the use of chemicals and their potential for both beneficial and harmful consequences, and altered the factors involved in the management of these resources.

The point that we should seriously contemplate, since our aquatic environment is quite susceptible to subtle changes, is that we must have an integrated system to evaluate the environmental impacts on fish and wildlife and the total effects on the nature of human existence.

During the past three decades, the Office of River Basin Studies and its succeeding organizations have coordinated efforts of the federal government to conserve and protect fish and wildlife resources of the United States as they relate to water resource development programs of federal and nonfederal scope. Conservation of these resources has prevented loss of or damage to fish and wildlife resources and provided for their development and improvement.

DIVISION OF FISHERIES RESEARCH

The ideal program for control of pest species involves an integrated system of careful pesticide usage, biological control, and physical manipulation of the aquatic habitat. Too often, chemical control has been used as a quick and easy solution to the problem, and this is usually to the detriment of fish and other aquatic life. We need to understand the ecological changes caused by herbicides, as well as the acute and chronic toxicity of these chemicals to fish and fish food organisms. Thus, our studies also involve food chain relationships and the fate, persistence, and toxicological significance of residues for the herbicide, its metabolites, and its interaction with other chemicals in the aquatic ecosystem. Biologists in weed control and fishery management meet on common ground in their mutual concern for an evironment free of pestilence and with good fishing, hunting, and aesthetic quality for our recreational enjoyment.

Our interests in management of this aquatic ecosystem also converge on the control of dense growths of aquatic plants that are responsible for the clogging of fishing and boating access, and on elimination of the protective habitat for mosquito larvae. This excessive plant growth affords cover that upsets the predator-prey relationship, resulting in stunted fish population and poor fishing success.[15] People using the area or living nearby complain about the insect problem, and, of course, the threat of an outbreak of malaria or encephalitis may exist. The fisherman may be just as intolerant of the situation, but has probably long since given up fishing because of the poor angling success. Nobody is happy at this point, and very often a drastic step must be taken to correct the situation. It is too late for application of management in the proper sense; we often treat the symptoms (spray for mosquitos, eliminate the vegetation and the fishery) and do not deal intelligently with the problem. We need better coordination of the various interests and agencies in these "fire-fighting tactics." To do this would require long-range planning with primary consideration given to environmental quality rather than "putting out the fire quickly and as cheaply as possible." Emphasis should be placed on safety and environmental quality as primary factors in decisions on whether or not to use a particular herbicide. We must insist that herbicides are used according to the label, with careful calculation of dosage, precision in application, and precautions

*From Walker, C. R., Chief Research Scientist, Division of River Basin Activities, Department of the Interior, Washington, D.C., Mimeograph Report, 1967.

for avoiding possible water or food contamination. Changes from the zero tolerance concept to finite tolerance limits present unique problems for the clearance and registration of pesticides used in or near aquatic sites.[16]

Impacts on Water Management

Not all bodies of water are easily managed, nor do their problems have simple solutions. Quite the contrary, we have a great deal to learn about aquatic ecosystems and how to properly manage them. Currently, no aquatic herbicide is registered that meets the criteria for all uses needed. Even the herbicides that may be necessary to manipulate plant populations in a body of water have limitations. Thus, we find many situations are difficult to deal with in many areas of this country. Since some herbicides are used directly in water for control of aquatic plants to enhance fish production and sport fishery, our investigations center on the fate of herbicide residues and effects on fish, fish food organisms, and other aquatic organisms.

Impacts on Aquatic Plant Control

Excessive aquatic plant growth is a universal problem in the management, culture, and harvest of fish. The role of aquatic flora in relation to the fishery is often poorly documented and not well understood. Generally, most research on weeds and weed control in fisheries has been focused on the removal of vegetation by chemical, mechanical, or biological means. Very little effort, however, has been made to understand the flow of plant nutrients and energy in the aquatic ecosystems and how selective biocides can be used to manipulate the species of producer organisms and food chain organisms to maximize the harvestable yield of fish. Thus, we must be concerned with basic research of aquatic ecology, plant and animal physiology, kinetics of plant nutrients in water and soil chemistry, toxicology of pesticides, nutritional biochemistry, life history of the aquatic food chain species (consumer organisms), and dependent relationships with aquatic plants and algae (producer organisms).

Herbicides are short-lived, but have a subtle effect on the aquatic ecosystem. The aquatic plants or primary producer organisms are directly affected, as is the objective of the management biologist. The changes induced, however, are transferred all the way up the food chain and dramatically alter the flow of energy. For example, sodium endothall is selectively toxic to certain submersed rooted plants and eliminates them from the habitat, releasing these stored nutrients and energy to decomposer organisms (bacteria, etc.), which in turn feed diatoms, rotifers, and protozoans. Turbidity from the growth of plankton is sharply increased but does not adversely affect feeding by predator-size fish at the secondary and tertiary trophic level. The net result is a more efficient system for benefiting the desirable sport fishery. Removal of excessive plant growth redirects energy flow and improves fish growth rates; increases in production and catch per unit effort are evidence of improvement in the sport fishery.

Herbicide evaluation requires an orderly system of toxicological screening and evaluation. Since herbicides may be applied in both standing (lentic) and flowing (lotic) situations, the bioassay methods must be sufficient to measure the concentration of herbicide, the contact time necessary for control of aquatic plants, and the toxicity to other aquatic organisms. Research methods should include both static and intermittent flow or constant flow bioassay systems, depending on the length of the testing and the investigator's desire to more nearly simulate the lentic or lotic environmental conditions. Temperature, ratio of biomass to volume, water chemistry, and light intensity or periodicity are important considerations with regard to the reaction of aquatic organisms or plants to the chemicals.[17]

Detailed studies are required to determine safe, practical methods of aquatic plant control for fish and wildlife management. Adequate labeling for safe and effective use of herbicides requires data to include the following: (1) toxicity to the target species; (2) relative toxicity to nontarget species, i.e., plants, invertebrates, fish, other aquatic animals, birds, mammals, and, by inference, man; (3) fate of residues and significance of residues in water, fish, crops, livestock, and foods; (4) condition affecting toxicity, efficacy, and persistence of residues in the proposed pattern of use, e.g., water chemistry, temperature, variations in susceptibility of species at various life stages and seasons, inflow dilution, contact time, rate of degradation, deactivation, or detoxification; and (5) synergizing or antagonizing activity of carriers, formulations, or combinations with other contaminants and pesticides, metabolites, and degradation products.

CORPS OF ENGINEERS
IMPACT ASSESSMENT

The science of predictive ecology is in its infancy. However, the environmental issues have been specifically supported by law under the National Environment Policy Act of 1969. The Corps of Engineers, as most agencies, has taken steps to determine what we do know, what we need to know, and to learn how to obtain information concerning what we do not know. Environmental knowledge will continue to expand, but when a decision is made, it must be made on the best knowledge available at the time.[18]

The Corps of Engineers has in progress several promising approaches to obtaining environmental information. The system is hierarchical and starts with an environmental reconnaissance inventory. This inventory contains a compilation of known environmental information regarding resources and amenities of regional or national significance. It is obtained from individuals, organizations, and governmental agencies. It includes features or species that are unique, rare and endangered, or that need to be treated with care. Where this inventory has been well prepared, neither the public nor the Corps of Engineers should be surprised by the introduction of new and important environmental information at late stages in the consideration of a project. These inventories are usually accomplished in conjunction with an investigation or a proposed project. This type of inventory will lead to, and provide information for, an environmental assessment.

Also related is a system that will identify needed baseline information for monitoring changes during project operations. Monitoring should tell if what is happening on a project is what is planned or anticipated, and, if things are going wrong, give accurate signals so that corrective action can be undertaken.

While this hierarchical information system is expected to provide important environmental information, much additional effort is going into research. Expertise is needed in dealing with any technical field and the environmental area is no exception. Orientation of staff members to environmental concerns has helped to ease the handling of environmental information. Both studies and advice are needed in support of investigations, and these are obtained as appropriate from individuals, universities, consulting firms, federal laboratories, and other governmental agencies in much the same manner as information required in the engineering or economic fields.

An important use of environmental information is in the preparation of an assessment of the environmental effects of a project, which is often given in an Environmental Impact Statement (EIS), as required by Section 102(2)(C) of the National Environmental Policy Act of 1969 (NEPA). Among the more apparent values of the Environmental Impact Statement are the requirements that the environmental aspects of a proposed action be evaluated in a regularized manner, and that this evaluation be made available locally and nationally to all interested parties.

Program of Impact Assessment

At this point it might be well to consider the evolution of the Environmental Impact Statement. The effect of NEPA on federal agencies has been dramatic, and extensive modifications in point of view and procedures have been evident. No other environmental legislation or federal guidance has had near the impact. In the Corps of Engineers alone, some 1200 impact statements were prepared and filed during Phase I.

Phase I of this program lasted about sixteen months after passage of the Act, or from January 1970 through April 1971. At an early stage of this phase it was accepted that almost all actions in the Civil Works program of the Corps of Engineers such as navigation, flood control, and shore protection projects, have major impacts on the environment and should be considered "significant" in the sense of the Act. This simplified the question of whether or not a particular action required preparation of an impact statement. It was also decided at an early stage that for purposes of preparing impact statements, the term "environment" should be understood to mean the natural environment.

Phase II began in May of 1971 with the updating of the various guidelines and regulations promulgated by the Corps to implement NEPA. The question of retroactivity of NEPA had become an area of great dispute and had to be addressed. This was the question of the applicability of the Act to projects that were under construction at the time the Act was passed. The number of projects either completed or under construction at that time was large. The Corps agreed, however, that taking the position that NEPA did not apply to these projects was not consistent with the immense complexities of environmental protection. As a matter of policy, an EIS would be prepared on each project in either a continuing construction or acquisition status and on each completed project in operation or requiring maintenance. In light of the workload involved, it was decided that these impact statements be prepared over a 3-year period. Priority was, of course, to be given to those actions about to be started and those which would have the most significant impact on the environment or entail an irrevocable commitment of resources.

Phase III began with a revision of the regulation on impact statement preparation. First, a schedule was prepared and published for annual review by the public and governmental agencies of all impact statements to be undertaken in the ensuing 3 years. The priority system and the policy that all significant actions will be supported by an EIS remained in force. What it said, in effect, was that projects with great environmental impact, or those where the environmental impact is controversial, would be accorded the highest priority. Second, it was recognized that some smaller actions can be environmentally insignificant and many are non-controversial. The reporting officer would be permitted to make an assessment, as if he were preparing for an environmental impact statement, and if the action appears to have little environmental significance, and this is not challenged by the public or governmental agencies, the file will so indicate, and no impact statement would be prepared. Third, perhaps of most significance, is that officers were encouraged to attempt to consolidate actions for which impact statements would be required into groups.

Most important, perhaps, would be the opportunity to examine a system as a whole, such as the locks and dams on the Ohio River, or all the reservoirs in a river basin, or the dredging activities along a long stretch of the intracoastal waterway. Statements can always be supplemented, augmented, or revised, and in the preparation of the environmental statements, specific areas and situations can be identified where additional study or refinement is needed.

The foregoing somewhat begs the question of what are the environmental facts. In part, this is because of the dynamic status of environmental information. We are learning more each day about

*Rivers and Harbors Act of 1970, Sec. 122.

what is critical and what may not be. The environmental facts that are sought are those that can influence the decisions to be made. These are acknowledged to be increasing in number and complexity. We are approaching the problem by improving the systems and procedures for bringing what is important into focus at appropriate points in the decision-making process.

Environmental Quality

Progressing to the second question, we now ask, what else is involved besides the environmental facts? Shortly after environmental impact statements began being filed in quantity, it became apparent that more information than was contained in the statement (EIS) was needed to make a decision on an individual action. In the summer of 1971, the Council on Environmental Quality began urging the incorporation of economic and other data in or with the impact statements. It was not disputed that more than environmental information was needed for making decisions. What was disputed was the identity of the decision document. It was accepted by that time that an EIS should provide "environmental full disclosure." As such, it supplemented the fundamental decision document or the report (the usual project report or other action document) prepared in almost every case involving a project's construction and/or operation.

The project report is usually addressed to the Congress and contains recommendations concerning the action proposed. Reports can be somewhat voluminous and, while available to the public, they do not have the visibility or readability of an impact statement. Extracts of the economics from the report and, later, another section of the report termed the Statement of Findings, were appended to the impact statement. This statement appears in a report immediately before the conclusions and recommendations, and in it the reporting officer rationalizes his findings, explaining how he arrived at the conclusions and recommendations in the overall public interest. In addition to the environmental considerations are economic, social engineering, and institutional considerations, as well as national policies and limits on the federal interest in any action.

The Congress recognized the increasing importance of a broader consideration of impacts, and in an Act,* about a year after passage of NEPA, required the Corps to deliver guidelines for the assessment of economic, social, and environmental effects of Civil Works Projects. As required by this law, the Corps of Engineers sent proposed guidelines to the Congress in June of 1972, and after review by other federal agencies, the states, and others, completed revision of the guidelines in 1972 for promulgation to all Corps of Engineers offices.[19] The assessments are similar in the case of the environmental effects to those required for the impact statement. In fact, completion of the assessment should enable preparation of the impact statement per se as a *pro forma* exercise largely involving a rearrangement of information. The intent is not to downplay the environmental consideration, but to highlight the full range of effects of any action.

Effect assessment considers the significant, hitherto examined, unevaluated, and unintended effects of any proposed action. These are effects that could not be considered in the past because there was no provision in existing procedures for their inclusion in the analysis.*

What is important is that the new effect assessment procedures make explicit what was stated as policy, as early as June of 1970, and in general guidelines[20] later that year, that environmental consideration must be brought to bear in every phase of Corps activity, from planning, through construction, to maintenance and operation. This is particularly important to the planning phase. The iterative assessment procedure is the same, of course, for the other effects as well as the environmental.

Worth mentioning is that while the Act speaks of economic, social, and environmental effects, the effects are not to be construed as mutually exclusive. Rather, the full breadth of possible impact or effect is considered for each effect identified. Each effect is assessed in each of the named categories (economic, social, and environmental) for convenience in communication, and to show interrelationship and mutual casuality. No significant effect (significant in terms of affecting a decision) should be excluded because of awkwardness in term fitting. For the purpose of assessment, the three terms — economic, social, and environmental — are defined so as to encompass the totality of human experience, substantially beyond fish and wildlife management considerations.

DIVISION OF RIVER BASIN STUDIES

The original Division of River Basin Studies of the Bureau of Sport Fisheries and Wildlife carried out the responsibilities of the Fish and Wildlife Service and the Secretary of the Interior under the Fish and Wildlife Coordination Act (48 Stat. 401, as amended; 16 U.S.C. 661 *et seq.*).

The Fish and Wildlife Coordination Act calls for fish and wildlife conservation to receive equal consideration and to be coordinated with other features of water resource development programs. It requires that agencies proposing water development projects for construction by the federal government and agencies considering applications for projects to be constructed under federal permit or license first must consult with the Fish and Wildlife Service and with the head of the concerned state fish and game agency, with a view to the conservation of fish and wildlife resources by preventing loss or damage and providing for their development and improvement.

In carrying out this responsibility of the Fish and Wildlife Service, the Division of River Basin Studies participates in the planning and review of projects of the Bureau of Reclamation, Corps of Engineers, Bureau of Indian Affairs, International Joint Commission (United States and Canada) and International Boundary and Water Commission (United States and Mexico), Alaska Power Administration; nonfederal hydroelectric projects subject to Federal Power Commission license; nonfederal, nuclear-fueled, electric power projects subject to Atomic Energy Commission license; nonfederal projects subject to Forest Service, Bureau of Land Management, or Department of the Army permits; and private reclamation projects under the Small Reclamation Projects Act. In addition, the Service participates with the Soil Conservation Service in planning federally assisted small watershed projects under provisions of the Watershed Protection and Flood Preservation Act (P.L. 83-566, as amended). By interagency agree-

*Benefit-cost analysis is the procedure by which an evaluation is made of the worth of proposed "public works." Benefits of the proposed action are compared with the costs of constructing or otherwise accomplishing the action. This procedure is rather standardized within any agency, but is also rather narrow. The ability of the procedure to reflect other than conventional economic efficiency values is limited. Multi-objective planning attempted by the Corps in 1966 on two major studies attempted to rectify this shortcoming.

ment it reviews water development proposals of the Tennessee Valley Authority, although the work of the Authority is specifically exempted from provisions of the Fish and Wildlife Coordination Act (FWCA).

Environmental Policy

The National Environmental Policy Act of 1969 directs all agencies to adopt and implement procedures designed to assure that their actions are in harmony with national environmental goals. By provisions of FWCA, the Division's delegated responsibilities for investigation of water development projects basically concern the overall environmental aspects of the projects with particular emphasis on fish and wildlife resources. Thus, the Division, by reason of its delegated responsibility and related expertise, has additional responsibilities under EPA for advisory service and review of environmental impact statements prepared under Section 102 of NEPA.

The reports prepared by the Division on federal water use projects are by law incorporated into the reports of the federal construction agencies and are available to the Congress when considering whether projects should be authorized for construction. Reports on nonfederal projects are transmitted to the federal licensing agency through appropriate departmental channels. These River Basin Studies reports recommend appropriate measures for incorporation in project plans for the conservation, development, and improvement of fish and wildlife resources. The Fish and Wildlife Coordination Act provides that the fish and wildlife measures which are acceptable to the construction agency shall be included as integral parts of the project plan and that the costs of such measures shall be included as a part of the project costs. On rare occasions, where a project could cause irreplaceable loss, the River Basin Studies report may recommend against construction of the project.

Fish and Wildlife Coordination

The original Division of River Basin Studies worked closely with the project construction agencies and with the state fish and game departments in developing fish and wildlife measures for inclusion in the projects. It worked closely with the National Marine Fisheries Service in determining effects of projects on commercial fishery resources and in developing measures for inclusion

in project plans to conserve, develop, and improve such resources. It also coordinated planning and estimates of benefits with those developed by the Bureau of Outdoor Recreation for other types of recreation.

The division was first established as the Office of River Basin Studies of the Fish and Wildlife Service in 1945. Shortly thereafter, the Act of August 14, 1946 (P.L. 732 — 79th Congress) amended the original Coordination Act of March 10, 1934. The amended Act provided much broader and clearer authority for the Secretary and the Service to participate in planning for fish and wildlife resources on water development projects and to recommend means and measures for the prevention of damage to fish and wildlife and for conservation and enhancement of these resources.

Prior to the time the Office of River Basin Studies was established, the consideration given to fish and wildlife needs in the water development programs of the nation had been piecemeal and largely conducted as a part-time effort by biologists employed for other duties. There was a huge backlog of projects on which investigations and planning had been completed by the construction agencies and on which fish and wildlife studies and planning were needed. Also, these agencies were investigating new projects each year which required study and planning for fish and wildlife. Little information was available as to effects of water development projects on fish and wildlife resources.

Shortly after the Fish and Wildlife Service was reorganized in 1956, the Office of River Basin Studies was succeeded by the Branch of River Basin Studies in the Bureau of Sport Fisheries and Wildlife, and in 1964, the Branch was raised to Division status. In the earlier years the River Basin Studies were provided with a very small annual appropriation and were dependent for the balance of funds upon field level transfers from the federal construction agencies, mostly on a project-by-project basis. In fiscal 1960 earmarked funds were included for this function not only in the Department of the Interior appropriation but also in the Public Works appropriations under both the Corps of Engineers and the Bureau of Reclamation. This improved financial structure has enabled the Division of River Basin Studies to plan its work more effectively. Funds available for Division activities from all sources now total about $5,800,000 annually.

Water Resource Projects

Since its establishment in 1945, the Division has prepared over 11,000 reports on water resource projects. Values or benefits assigned to fish and wildlife in these reports have been included by the federal construction agencies in the economic justification for water projects and used in allocating project costs. Fish and wildlife measures recommended in the reports have been incorporated into a number of projects now completed and in operation.

The Division is actively involved in the Corps of Engineers navigation permit program covering both dredge and fill activities under Section 10 of the Act of March 3, 1899, and refuse (effluent) disposal under Section 13 of the Act. If investigations under the permit program indicate that the work planned would damage fish and wildlife, recommendations are made to modify the plants to prevent the damage. If no acceptable plan can be developed, denial of the permit is recommended. A 1967 Memorandum of Understanding between the Secretary of the Interior and the Secretary of the Army delineates the procedure for handling difficult permit cases. Some of the more difficult cases are referred to the Secretarial level for resolution.

The majority of River Basin Studies reports have been on projects still to be authorized or licensed for subsequent construction. Consequently, the Division has been building into the plans for future water developments, provisions for fish and wildlife which will meet the needs of future generations. An appreciation of the scope of the program can be gained from an accounting of the land acreages turned over to the states at the few projects underway or completed through 1970. A total of 811,831 acres of federal land has been turned over to 38 states for wildlife management at 140 water development projects. Of this acreage, 536,323 acres (66%) are mitigation lands, i.e., lands set aside in partial compensation for project-caused losses, and 275,508 acres (34%) are enhancement lands or lands set aside for improvement of habitat, beyond or in absence of any project-caused losses.

Among other measures which are recommended for incorporation into projects are minimum releases from reservoirs to improve downstream fisheries, minimum pools in reservoirs to sustain the reservoir fishery, protective devices to prevent fish from entering turbine intakes or irrigation canals, passage facilities for upstream and downstream migrant fishes, fish production facilities where passage is infeasible, selective reservoir clearing to improve utilization by fishermen, creation of subimpoundments within reservoirs for fish and wildlife management, and acquisition and development of project lands for establishment of national wildlife refuges. Plans for fish and wildlife measures are coordinated closely with the construction agencies to assure that fish and wildlife conservation is considered at the proper time in project formulation and construction.

National Wildlife Refuges

Among the numerous recommendations of the Division which have been accepted and built into water use projects are several related to maintenance of anadromous fish. These include Nimbus Fish Hatchery at the Nimbus Reservoir project on the American River, Trinity Fish Hatchery below Lewiston Dam on the Trinity River, and an artificial spawning channel below the Red Bluff Diversion Dam on Sacramento River, all features of the Bureau of Reclamation's Central Valley Project in California. Several other anadromous fish hatcheries were built earlier, around 1940, by the Bureau of Reclamation at the Columbia River Basin and Central Valley projects. The Corps of Engineers also has built several anadromous fish hatcheries as well as many major fishways and related facilities in the Columbia River Basin, one of the latest being the world's largest steelhead trout hatchery at the Dworshak Dam in Idaho. Listed in Table 3.1 are national wildlife refuges which have been built into water development projects as a result of River Basin Studies recommendations.

The Garrison irrigation project (authorized in 1965) in North Dakota and South Dakota includes plans for 36 major fish and wildlife development areas and a number of minor ones. These developments typically involve dikes with control structures across low areas and the provision of water from project canals to form or improve waterfowl marshes. Excellent habitat will be available in both wet and dry years for waterfowl and other wildlife. Hunting and fishing opportunities will also be increased significantly. Administration of the areas will be divided as follows: 26 by the Bureau of Sport Fisheries and Wildlife, 8 by the North Dakota Game and Fish Department, and 2 by the South Dakota Department of Game, Fish and Parks.

These fish and wildlife developments involve

17

TABLE 3.1

Water Development Projects Containing National Wildlife Refuges

Name (NWR)	State	Project	Constructing agency
Choctau	Alabama	Jackson Lock and Dam	Corps of Engineers
Eufaula	Georgia and Alabama	Walter F. George Lock and Dam	Corps of Engineers
Kirwin	Kansas	Kirwin Unit	Bureau of Reclamation
Washita	Oklahoma	Foss	Bureau of Reclamation
Flint Hills	Kansas	Jackson Lock and Dam	Corps of Engineers
Umatilla	Washington and Oregon	John Day Lock and Dam	Corps of Engineers
Big Stone	Minnesota and South Dakota	Big Stone Lake – Whetstone River Reservoir	Corps of Engineers
McNary	Washington	McNary Lock and Dam	Corps of Engineers
Seedskadee	Wyoming	Colorado River Storage	Bureau of Reclamation
Audubon	North Dakota	Garrison Reservoir	Corps of Engineers
Cross Creek	Tennessee	Barkley Lake and Dam	Corps of Engineers
Cibola	Arizona and California	Colorado River Front Work and Levee System	Bureau of Reclamation
Sequoyah	Oklahoma	Robert S. Kerr Lock and Dam	Corps of Engineers

146,530 acres of land and a related supply of 165,123 acre-ft of water during dry years. The federal cost of these developments based on 1970 price levels is expected to amount to $23,135,000. Preauthorization planning work is now programed for the ultimate stage of the Garrison Diversion Unit, which involves an additional 750,000 acres of irrigable land and some 35 additional fish and wildlife development areas.

A good example of what can be accomplished with a smaller project is found on the authorized Narrows unit located on the South Platte River in Colorado. Fish and wildlife facilities of the project include rehabilitating a 2000-acre storage reservoir for general recreation and fish and wildlife, establishing a 15,765-acre state wildlife management area at the upper end of Narrows Reservoir, and construction of a fish hatchery below the dam. Total fish and wildlife costs are estimated at $3,575,000. Besides compensating for the fish and wildlife losses attributed to construction of Narrows, these facilities will provide the additional benefits of 144,000 man-days of fishing, 1500 man-days of upland game hunting, 6000 man-days of waterfowl hunting, and 23,000 man-days of wildlife-oriented recreation.

Nonfederal Projects

The Division prepares letter-form reports including suggested license stipulations and other language for inclusion in the Department's report to the Federal Power Commission on nonfederal

hydroelectric power projects for the purpose of conservation and development of fish and wildlife resources. In most cases, the Federal Power Commission incorporates the Department's recommended stipulations in permits and licenses which it issues for private power projects. These efforts have resulted in substantial investments in facilities for fish and wildlife. In 1968, private and public utilities licensed by the Federal Power Commission reported investments up to that time totaling over $116,500,000 in facilities specifically for fish and wildlife and annual expenditures of over $1,100,000 for the operation of these facilities. At the Roanoke Rapids and Gaston projects, located on the Roanoke River in North Carolina and Virginia, studies conducted by and in cooperation with the Division indicated that substantial downstream flows and water quality control devices were needed at the projects for protection of the spawning area of an important population of striped bass. Underwater weirs were constructed at both projects so that the higher quality surface water is discharged downstream and a minimum release is provided to maintain conditions for striped bass spawning in the lower river.

As a result of the Division's report on the existing Leaburg project, located on the Mac-Kenzie River in Oregon, it was recommended that protective screens be installed to prevent the destruction of anadromous fish passing the project. A horizontal traveling screen is being constructed which will reduce fish losses substantially.

During the review of plans for the Oroville project, located on the Feather River in California, the Division reported that a hatchery was needed to maintain runs of anadromous fish and a multilevel outlet was needed to control the temperature of water released to the Feather River downstream. Through subsequent actions before and by the Federal Power Commission, a temperature control device was installed to provide suitable temperatures for both cold water and warm water fish species as well as for other purposes, and a fish hatchery was installed to maintain anadromous fish runs in the river. The costs for construction of these facilities were about $12,000,000 for the multilevel outlet and about $3,655,000 for the hatchery.

Under a Memorandum of Agreement between the Atomic Energy Commission and the Department of the Interior, the Division coordinates reports of its investigating staff and those of the National Marine Fisheries Service through their radiological laboratory and prepares the Bureau's reports to AEC on the effects nuclear plants would have on fish and wildlife resources. These reports include recommendations for radiological and ecological monitoring programs, and measures to protect the fish and wildlife resources and the environment from the effects of radiological and nonradiological wastes including thermal, chemical, and other effects.

The Commission requires the licensee to provide evidence that the project will pose no threat to human health and safety and requires that the licensee comply with all state and local laws concerning land use and water quality. The Commission also includes the Bureau's report in the record of project proceeding, thus making the report of the license.

Power companies, in recognition of the increased public concern for the environment, have adopted most of the Bureau's recommendations. Radiological and ecological monitoring programs are being included or are planned for all projects. Many companies have installed or plan to install cooling facilities to limit thermal effluents so as to meet water quality standards and protect aquatic resources.

Watershed Protection and Flood Prevention

The Division of River Basin Studies also carries out the responsibilities of the Secretary of the Interior under Section 12 and other pertinent sections of the Watershed Protection and Flood Prevention Act. Section 12 authorizes the Secretary at his choosing to conduct fish and wildlife studies of small watershed projects, but it limits acceptance of recommended measures to protect and enhance fish and wildlife to those sanctioned by the local sponsors and the Secretary of Agriculture. The Division studies such projects in close cooperation with the appropriate state fish and game departments and recommends measures for preventing damage and improving conditions for fish and wildlife. Section 4 of the Watershed Protection and Flood Prevention Act authorizes federal cost-sharing of fish and wildlife improvement measures incorporated in these projects and thereby provides incentives for state fish and game departments to cooperate in the program.

The Federal Water Project Recreation Act (P.L. 89-72) increases the responsibilities of state fish and game agencies and other nonfederal interests in planning and funding the recreation and fish and wildlife enhancement features of federal water resource development projects. It calls for cost-sharing of certain separable features added specifically for recreation and fish and wildlife enhancement. It also provides for administration and operation of such facilities and areas by nonfederal public bodies. The Act, in effect, recognizes that both local and nationwide benefits flow from recreation and fish and wildlife enhancement features in water use projects. The Division works with prospective administering and cost-sharing agencies in developing plans which are within the range of their financial capabilities and interests.

Water Resource Planning

The work of the Inter-Agency Committee has been supplanted by more formal arrangements under the Water Resources Planning Act of 1965. That Act provides for the coordination of federal policies and programs with those of all levels of government connected with the water resources field. The Act has three main objectives: (1) to facilitate coordination of Federal policy by establishing a Water Resources Council comprised of the heads of the departments having major concerns with water resources planning; (2) to foster coordination of governmental activities within individual river basins by authorizing commissions made up of representatives of interested federal agencies, states, and interstate and international agencies to prepare and maintain "a comprehensive coordinated joint plan"; and (3) to

19

strengthen the planning activities at the state level of government. The Division has become increasingly involved in comprehensive water planning under activities led by the Council. These include the comprehensive basin studies being conducted nationwide, the work of established Regional River Basin Commissions, and the first National Water Assessment (1968).

The Division sponsored in 1969 a number of conferences in an effort to strengthen its coordinative position with the state fish and game departments and with national conservation organizations. Representatives of virtually all of these organizations attended a National Symposium in October 1970, or one of the four follow-up regional workshops. The objective was to air all of the known policy, procedural, and legislative problems and to arrive at recommended solutions. The result was an apparent increase in appreciation for the importance of the work and a new determination to develop a unity of purpose and action to make the national water resources development program better serve the interests of fish and wildlife conservation. New and amended policies and legislation were recommended and a number will be implemented as soon as practicable.

Participation in the drafting, amending, and review of federal legislation is an important Division activity. The Division reviews and provides comments each session of the Congress on numerous bills which pertain to water resource planning in which the Service has an interest. The Division also furnishes factual information on fish and wildlife resources and related material, particularly in relation to water and land development programs, for use of Senate and House Committees. The Division prepares testimony materials for presentation at Senate or House Committee hearings and its personnel serve as technical witnesses when required at such hearings.

One outstanding example of River Basin Studies participation in legislative affairs was in connection with passage of the strengthening amendments to the Fish and Wildlife Coordination Act of 1958. The Division drafted language for the amending bills in cooperation with the Department's Office of Legislative Counsel, contributed materials on the proposal for submission to State Governors, and developed and presented testimony to Congressional committees on the measure.

The end product of the Division's principal efforts in the legislative field is realized when acts authorizing water development projects also authorize measures needed for fish and wildlife conservation. This may be accomplished either by language in the acts themselves or in the legislative history of these acts, including service reports which are integral parts of project documents.

Fish and Wildlife Resources

The Division of River Basin Studies has had primary responsibility for carrying out several major studies related to fish and wildlife resources. The National Estuary Study responsive to the requirements of the Estuary Protection Act (P.L. 90-454) is a recent example. This entailed a comprehensive study of the nation's estuarine environment, the uses of estuarine resources and the resulting conflicts, and evaluation of legislative and institutional mechanisms, to provide a basis for needed public actions to safeguard natural values of such areas, while allowing for reasonable and compatible economic development.

The Colorado River Basin Project Act of 1968 provided for a survey of the water resources of the 11 western states which lie wholly or partly west of the Continental Divide. This Act directed the Secretary of the Interior to conduct a comprehensive reconnaissance investigation and develop a plan called the Western United States Water Plan Study (Westwide Study) to meet the current and projected water needs of the 11 western states. The Bureau of Sport Fisheries and Wildlife is pursuing an active role in the Westwide Study. It views this study as an opportunity to participate as an equal partner in planning that will shape the future along desirable lines.

The National Surveys of Fishing and Hunting, covering the years 1955 to 1970 were conducted under direction of the Division. They are based on sample surveys of the United States population to determine such factors as participation rates in hunting and fishing, expenditures of time and money in such activities, and other relevant statistics. The Division spearheaded the collection of fish and wildlife data, by major river basins, for the National Water Assessment. The data and text were designed to indicate the close relationship between water and fish and wildlife, the demands placed on these resources by people for recreational and commercial uses, and the probable future trends and requirements.

Chapter 4

HYDROPOWER PROJECTS AND THEIR
ASSOCIATED ENVIRONMENTAL IMPACTS*

INTRODUCTION

Although the energy crisis and concern for the environment seem to have come upon us recently, this crisis has been in the making for some period of time. We observe that hydropower is the cleanest energy in the world, even though it is not without environmental problems. Environmental considerations can be grouped into five resource areas, namely, geological, land, water, biological, and cultural. The systematical analysis of these resources leads us to a reordering of priorities in the development and planning of projects. As a public agency, the Corps of Engineers must be responsive to the public interest, but in this process, there must be an equal analysis of the tactics that are being used to prohibit the development of these resources which are indeed a necessary part of the civilized government of our society.

The role and importance of hydropower is growing in the United States and in many other countries[39] since fossil fuels — oil, coal, and natural gas — are limited. We are relying on foreign producers of these fuels to meet much of our need. Dependence on foreign energy sources contributes to economic uncertainty and instability, and the nation is now searching for alternative sources of power.

Investigations are being made as to the reliability and potential for utilizing renewable energy resources[25,26] other than hydropower, such as solar or geothermal sources for producing electricity. Many new technologies are currently being examined on methods of heating and cooling buildings with solar energy. The principal impediment to solar energy is not technology but the cost of materials used in solar heating and cooling, and the reluctance to invest in untried and unproven systems. Electrical energy from geothermal sources show some promise.[24] We have one commercial geothermal electric generating facility in the U.S. in the geyser area of northern California. The federal program for geothermal developments is limited, however, to exploration, advanced research, and technology development.

Any real contribution to the nation's needs for electricity to come from geothermal sources is many years in the future. As one examines the immediate and foreseeable need for renewable sources for energy, hydropower remains the most feasible and economical[21,32] (see Figure 4.1).

A hydroelectric project harnesses the potential energy of a river's gravitational fall to produce electricity. Water is stored behind a dam and released at desired intervals through conduits called penstocks or power intakes. The conduits direct the stored water to large turbines installed near the tailwater level of the dam. The force exerted by the water on the turbine blades drives the turbines, which, in turn, drive large generators to produce electricity.

POWER DEVELOPMENT IN THE UNITED STATES

The United States has a total capacity of over 524 million kW by the combined production of hydropower and thermopower generation and produced over 2 trillion kWh of electricity in 1976 (see Table 4.1.). Most of our electricity comes from thermopower methods using fossil and nuclear fuels. Hydropower techniques are used, however, to produce a significant quantity of the electricity required by our nation, over 303 billion kWh in 1976, and this comes from a renewable energy source, water. Nonfederal agencies (cities, states, private companies, etc.) produce over half of the hydropower produced electricity of this country with a capacity of over 38 million kW. The federal government controls and manages an additional capacity of nearly 28 million kW. The federal government's capacity includes conventional hydropower, as well as pump storage capabilities (see Table 4.2).

There are over 150 federal hydroelectric power plants in operation, seven with reversible equipment for pump storage or pump back capability. The pump storage-pump back concept provides a method for smoothing out the power demand curve. When you have excess power and low demand the power can be used to pump water

By C. G. Ash, Chief of Environmental Policy Civil Works, U.S. Army Corps of Engineers, Washington, D.C.

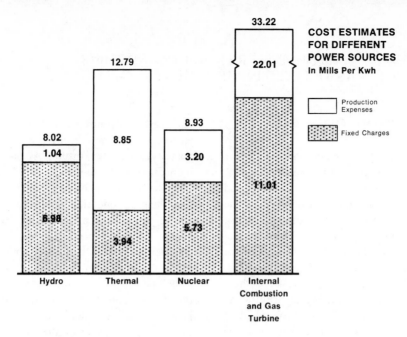

FIGURE 4.1. Lower production costs of hydropower are partially offset by higher construction charges. Installation costs per kilowatt of installed capacity of a hydroelectric plant average between $200 and $300. Nuclear plants average $175 per kilowatt, fossil fuel facilities $121, and internal combustion and gas-fired turbines $84. (From Department of the Army Corps of Engineers, EP 1165-2-3, July, 1976. With permission.)

TABLE 4.1

Total Electrical Power in the United States

(Hydropower, Fossil and Nuclear Fuels)

Capacity (total U.S.)	Kilowatts
Installed (hydro, fossil, nuclear)	524,200,000
Non-hydropower (fossil, nuclear)	457,900,000
Hydropower	66,300,000
Hydropower without pump storage	56,600,000
Hydropower pump storage only	9,700,000
Hydropower (nonfederal)	38,600,000
Hydropower (federal)	27,700,000
Production (total U.S.)	Kilowatthours
Installed (hydro, fossil, nuclear)	2,001,700,000,000
Thermopowered (fossil, nuclear)	1,698,400,000,000
Thermopowered (fossil)	1,526,500,000,000
Thermopowered (nuclear)	171,900,000,000
Hydropowered (14.6% of total)	303,300,000,000

Data from Federal Power Commission Annual Report, 1976.[37]

back into the main reservoir or some other container. When the demand for power is again increased the water can be reused to produce electricity. There are 18 new federal hydroelectric plants currently under construction, five of which will have the capability for pumping back water. All together the federal and private sectors produce about 303 billion kWh of electricity per year by hydropower methods (see Table 4.1). This seemingly astronomical figure is still only 14.6% of the total U.S. electric power production. Looking at it in another way, the hydropower production in the United States represents an annual savings of nonrenewable natural resources on an energy equivalent basis of 130 million tons of coal, 3.2 trillion cubic feet of natural gas, or 519 million barrels of oil.

HYDROPOWER DEVELOPMENT BY THE CORPS OF ENGINEERS

The Corps is the nation's largest single producer of hydroelectricity. As of January 1976, the Corps operates 65 projects housing 293 turbine-generator units with a total installed electrical capacity of 15.6 million kW (see Table 4.3). In 1975, Corps facilities produced 85.3 billion kWh of electrical

TABLE 4.2

Total Federal Hydropower in the United States

(Agency Installed Capacity)

	Conventional (kw)	Pump storage (kw)
Tennessee Valley Authority	3,150,000	60,000
U.S. Army Corps of Engineers	15,640,000	28,000
Bureau of Reclamation	8,150,000	565,000
Other	125,000	–
Total	27,065,000	653,000
Under Construction	7,230,000	2,070,000
Authorized	38,870,000	3,440,000

Data from Federal Power Commission Annual Report, 1976.[37]

energy. Corps hydroelectric energy production was 24% of the total U.S. hydroelectric energy production and 3.3% of all electrical energy produced in the nation that year. To produce that much power from the nation's nonrenewable sources would have required burning 35 million tons of coal, 800 billion cubic feet of natural gas or 5.5 billion gallons of oil. To express it one more way, the Corps of Engineers energy production was equivalent to the output of 20 average-sized nuclear plants.

The Corps is one federal agency that has a long history in the field of hydropower. The Corps of Engineers involvement in hydroelectric production stemmed from its growing responsibility for the nation's water resources. In 1824, Congress assigned the Corps the task of clearing snags and sandbars from the Ohio and Mississippi Rivers, and this initial assignment gradually expanded to a general responsibility for navigation improvements.

Then in 1909 the federal government acquired a dam on the St. Mary's River in Michigan. Though the acquisition was primarily for navigation purposes, the site also contained a hydroelectric plant, the Corps' first. Also that year, Congress recognized the potential significance of hydroelectricity to the growing nation and directed the Corps of Engineers to include assessments of water power potential in its periodic surveys of U.S. waterways. At that time, development of hydroelectric facilities was conducted almost entirely by private interests. The Corps of Engineers began hydropower construction in 1916 when it added a

TABLE 4.3

Hydropower Capacity and Production in Corps of Engineers Facilities

In operation	
No. or projects	65
Capacity in operation	15,641,195 kW
Capacity under construction	3,293,960 kW
Ultimate capacity	19,045,750 kW
(In operation) % of total U.S. hydro	24%
(In operation) % of total electrical capacity	3.3%
Hydroelectricity produced (1975)	85,374,193, 121 kW
Under construction	
No. of projects	6 (17 units)
Initial capacity	927,000 kW
Ultimate capacity	2,954,375 kW
Total in operation and under construction	
No. of projects	71
Generating capacity	16,957,475 kW
Ultimate installation planned	22,000,125 kW

Data from Federal Power Commission Annual Report, 1976.[37]

turbine unit to the St. Mary's plant. Particularly since the end of World War II, the Corps of Engineers role in hydroelectric development has greatly expanded and spread to all regions of the country. While many public and private groups and two federal agencies, the Bureau of Reclamation and the Tennessee Valley Authority, have also developed water power resources, the Corps of

Engineers is the nation's largest builder and operator of hydroelectric facilities (see Table 4.2).

ENVIRONMENTAL CONSIDERATIONS

Hydropower is essentially a product of solar energy. The sun evaporates water from land and sea, and this moisture collects in the atmosphere to later return as rain or snow. Much of the water falls on high lands and drains into tiny streams and large rivers as it flows back down to the sea. This flow, together with the difference in elevation, known as head, can be harnessed by a hydroelectric project to produce electricity.

Although the energy crisis and concern for the quality of the environment have sparked research into new methods of environmentally clean electrical power production, hydroelectric generation is the cleanest source of electrical energy available today. Thermal generating facilities, either nuclear-fueled or fossil-fueled, are now the only alternative sources capable of producing electrical energy in the quantities needed to support a modern industrial society, and there are significant environmental problems associated with technologically feasible means of thermal generation. All thermal generating facilities, for example, require water either for steam production or for cooling, and the discharge of the heated effluent into natural watercourses or reservoirs can create problems of thermal pollution. Fossil-fueled thermal plants have the additional disadvantages of consuming vast quantities of nonrenewable natural resources and either emitting pollutants into the air or requiring costly control devices to reduce emissions. Nuclear-fueled thermal plants suffer from the disadvantage of the possible uncertainty of assuring public health and safety in the vicinity of such plants and for handling and disposing of their radioactive wastes.

Hydroelectric projects also have environmental disadvantages, but the disadvantages are generally regarded as being more easily mitigated or more socially acceptable. The very construction of a dam and reservoir entails significant environmental consequences, some of which can be avoided by careful site selection, some of which can be mitigated through enlightened planning and management[29,33] of the projects, and some of which cannot be avoided or mitigated but must be accepted until we can develop alternative means of power production that have less severe consequences.

Hydroelectric power generation itself may produce adverse environmental effects beyond those inherent in the construction of dams and reservoirs, but engineers and scientists are making important advances in eliminating many of these effects. Technology is available for solving many of the adverse effects but they cannot be solved unless the environmental impacts are recognized.

Some environmental impacts can to a certain extent be quantified in terms of dollars, acres, or with numbers of one kind or another.[42,46] There are impacts, however, that defy quantitation and must be dealt with in a subjective manner. Many times we are restricted to judgment evaluations, based on the experienced eye of the observer, which may vary dramatically according to the background, interests, and objective of the individual. It is our objective, therefore, after examining a number of Corps of Engineers hydroelectric projects, to present an overview or combined analysis of representative projects across the country in the hopes of highlighting the more significant environmental opportunities, as well as some of the environmental concerns associated with hydroelectric projects. As a matter of convenience, we have quite arbitrarily separated environmental considerations into five categories: geological, land, water, biological, and cultural resources.

Geological Resources

In examining the geological characteristics of project areas there are several obvious adverse impacts. First, the dam site and reservoir will be permanently inundated. Certain natural formations will be covered with either water or earth and some of the natural scenic beauty will be lost. Perhaps the attractiveness will be replaced with a different kind of beauty, but one thing is certain the natural scene will be changed. While these aesthetic values are hard to quantify, they are nevertheless important considerations which must be considered in any water resource development.[28]

Other impacts do lend themselves to quantification; for example, certain mineral deposits, i.e. coal, bauxite, gravel, diamonds, etc., may be covered. These resources are not necessarily lost forever, but there will be a definite increase in the cost and difficulty associated with mining these materials. With a little effort and assistance from mining professionals, a fairly accurate impact can usually be determined and expressed in terms of actual costs.

Changes will develop in the ground water systems in the vicinity of the reservoir. Siesmic activities in the local area must be evaluated. Sometimes it is very difficult to obtain all the scientific information one would like on aquifer and siesmic phenomena. We are often left with only the good judgment of the experts in the field to form a basis for evaluation of these sometimes critical activities. Caves and springs may be covered and lost in the dam site and pool areas. New springs can develop below the dam. Caves which were once dry in the area may become wet. However, the impacts are not always adverse to the area. The new aquifer system may have a highly beneficial impact on the surrounding community; for example, the increased rate of recharge into the aquifers may provide a new source of water for man and animals.

Land Resources

The obvious adverse environmental impacts on land are the losses resulting from the creation of the dam and reservoir. The land areas completely inundated are usually a combination of land types — agriculture, riparian growth, hardwood forest, savanna or pasture, wetland marsh, etc. Certain aspects of these losses can be quantitated. For example, the value of agricultural produce or timber products can be measured in dollars and cents. The impacts on the fish and wildlife as a result of the changes are not so easily quantitated. Such items as species sought by fishermen and hunters can be evaluated, but for the most part we are left with some subjective judgments. The expert may have to make comparative evaluations of the advantages as well as the disadvantages associated with the changes and rely on his or her experience to project an assessment of the future situation.[21,22] In making this assessment, it is important to project what will likely take place as a result of natural and man-made causes, even if the project is not built, and compare these findings with what is projected to occur if the project is constructed. Only by this type of analysis can the decision maker be provided with an evaluation from which sound judgments can be made.[31,46]

Hydropower projects with their fluctuating demands for water give us a constantly changing shoreline for the reservoir. In some situations the water level can drop 10 to 20 ft in 8 to 12 hr. In the upper end of the pool where the water is shallow such a drop will in some instances leave a mud flat of several hundred acres in size. These muddy areas may last for weeks or months depending upon the seasonal rainfall and recharge capacity of the river basin. Mud flats are sometimes unsightly and the Corps of Engineers receives unfavorable comments from the local residents who assess the situation on aesthetic grounds. There are situations, on the other hand, where these lands can be used quite effectively for agricultural or wildlife conservation purposes and aesthetic difficulties can to a certain extent be ameliorated. Sometimes the more important impacts are more subtle. For example, certain fishes spawn in shallow waters. If the pool level drops quickly their eggs are left high and dry on the mud and are destroyed. This will certainly influence the numbers and types of fish which can propagate successfully in the reservoir. Understanding the life history requirements of the species desired and coordinating this knowledge with the operation of the facilities is essential to the management of a viable and healthy fishery in association with a hydroelectric power reservoir. This is true not only for the waters in the reservoir but also the riverine area below the dam.

Reservoirs trap sediments carried in from the upper basin rivers and streams. In Corps of Engineers projects we try to design the pool with sufficient capacity for 100 years of siltation deposits without causing a change in project operations. For the most part, our engineers have been very accurate in the forecast for the accumulation of sediments. However, this is not an exact science, and should the forecasted use of land in the upper watershed change from what was predicted, an increase in downstream sedimentation may occur. There have been a few projects where the sediment load has caused considerable concern. Fishtrap Lake and Dam,[34] completed in 1968, located in the rolling, rugged hills of Kentucky, is an example of a project where sediments are accumulating faster than predicted and at an alarming rate. The problem here is that no one at the time this project was designed could foresee the rapid expansion of strip mining for coal in this area. In the mining process the vegetation is cleared from the hillsides and the overburden soils and rock are stripped from the coal and left in loose piles. When it rains there is no vegetation to stop the loose materials from being washed into the streams and then carried into the reservoir. This problem is gradually being brought under control by restricting some areas from mining and building a number of dikes and settling ponds. But the new construction has been costly, time-consuming, and a source of controversy.

The manner and extent of the vegetation removal from the pool area during construction and before the reservoir is filled can have environmental consequences. Here there are perhaps two extremes: to remove all vegetation from the pool area versus the decision not to remove enough plant (trees and shrubs) material. The removal of all vegetation results in a clear, clean reservoir, aesthetically pleasing, and free of any obstacles which would hamper boating and other recreation on the lake. Such an action is very costly and in many cases the cost is prohibitive. Fish and wildlife managers often prefer that a reasonable amount of trees and other vegetation be left to provide protection areas and proper conditions for fish and other aquatic species to reproduce. Dead or dying trees that protrude or are barely below the water surface are not attractive and are hazardous to boats. During the first year after inundation, where there are copious amounts of vegetation to deal with, other problems can also be encountered. Floating debris is hazardous to boats and is aesthetically displeasing, to say nothing of the bad odors that may develop. Most of the woody materials become waterlogged and eventually settle to the bottom. Microorganisms which decompose the organic matter on the bottom of the pool utilize the oxygen in the water during the decomposition process, to the detriment of large organisms. The larger species require greater quantities of oxygen to survive and are therefore in a poor competitive position. The result can kill certain fish and crustacean species during the first few months or year after filling the pool.

In considering the environmental impacts of reservoir construction on the land resource, the adverse impacts associated with the relocation of towns and facilities which are flooded or inundated should not be overlooked. Nor should secondary impacts be overlooked, such as the tendency for people to locate new houses, farms, and businesses in the flood plain below the dam, attracted there by the feeling of protection from flooding. Such protection is not always supported by the design of the dam or the operating procedures.

There are also the beneficial aspects which must be weighed in accordance with the public's best interest. The still water reservoir with its added recreational facilities and opportunities will be desired by many people rather than a free flowing stream. Flood protection is a desirable feature and the benefits from the electricity produced are always in demand. The reservoir may also provide a supply of water for potable and other domestic purposes.

Water Resources

Let us examine the water resources more specifically and identify some of the adverse and beneficial impacts associated with the water.[36] We have earlier mentioned the conversion of a free flowing stream to still water and noted that there will be a change in the fishery. It has also been noted that the flow and dynamics of the stream below the dam will be modified from the natural condition. The condition and significance of the changes of the stream will depend to a great extent on the amount of water allowed to pass the dam. Water quality can differ depending on the amount of flow. For example, when a dam is built on a large stream such as the Columbia River, water continually flows over the dam, and air mixes with the water on the spillway. The spillway dumps into the stilling basin. The water in the stilling basin below our dams varies in depth up to 90 feet. The water plunging deep into the stilling basin is subjected to high pressures and the water becomes supersaturated with certain gases, e.g., oxygen, nitrogen, etc.[41,44] During high water periods on the Columbia River gas supersaturation measured in terms of nitrogen has on occasions reached 140%.

Fish differ in their responses to gas supersaturation. Salmon begin to show signs of stress from gas bubble disease at approximately 110% gas supersaturation.[23] The water in the fish's tissues becomes supersaturated. The fish encounters little or no stress if it stays in deep water, but as it comes to the surface the excess gas in the tissues expands causing bubbles to form in the tissue, similar to the bends in divers. If these bubbles form in a critical part of the body the fish dies. It is particularly severe in small fish. The Corps of Engineers has had some success in the control of supersaturation by the construction of a lip at the base of the spillways which forces the water out and downstream rather than plunging to the bottom of the stilling basin.

The water passing through the electric turbine does not generally supersaturate. When the flow of water over the dam is relatively small, or only intermittent, gas bubble disease of fish is of little consequence.

The water in many reservoirs tends to stratify into two or more temperature zones. The location of the thermocline boundaries between these zones and the physical conditions associated with these layers is dependent upon a number of factors: the depth of pool, the rate of exchange of water, inflow water characteristics, light penetration, air and soil temperatures, seasonal fluctuations, etc. Every reservoir has its own characteristics and conditions. Many of these situations were predictable and the projects were designed with built-in controls for management. Space will permit only a few examples of the effects of stratification on reservoir life and management.

The lower cooler zone tends to become anaerobic as a result of microbial decomposition of the organic matter continually deposited in this area of the reservoir. Because of reduced oxygen levels, life in this zone is limited. Water discharged through the outlet structures from this zone transmits to the stilling basin and river below the dam the same oxygen depleted conditions. Aerobic flora and fauna cannot survive until such a time in the streams' movements that the waters become aerated and the oxygen is replenished.

In some reservoir projects, multiple intake structures are built which permit a choice of from what depth the water is released. By choosing the shallower aerated water, anaerobic conditions can be prevented from occurring below the dam. Within limits, the water temperatures below the dam can also be controlled with these structures. The use of the cooler waters allows the establishment of a cold water fishery (trout) in areas where only warm water fishes were previously found.

The pH or acidity of the water is important.[2 3] In the Corps of Engineers there have been some dramatic experiences when reservoirs were filled with water that drained from coal mining areas. Acid mine drainage drops the pH of the water and in some instances most of the plant or animal species died. For all practical purposes a dead body of water can be established.

There is much more to water quality at reservoir sites than free oxygen levels, pH, and temperature. The water can be polluted with organic and inorganic substances, some of which may be toxic to aquatic life. Suspended solids and particulate matter, both living and dead, cause turbidity which can also be an important factor. Turbidity can be separated into two types, that caused by suspended solid materials originating from erosion of soil, and that caused by the growth of millions of tiny plankton which multiply in suspension with the water column. The operation of a facility can be important in controlling turbidity. Often the reservoir will be turbid as a result of bank erosion resulting from wave actions, or fluctuations in pool level caused by rapid drawdown of the water in the reservoir associated with electric power production. When this problem is recognized, the rate of discharge of water can usually be controlled without interfering with the generation of power.

Turbidity limits light penetration, which limits plankton growth. These small floating plants, animals, and microbes are the principal source of food for many of the higher forms of life in the reservoir. Many of the microorganisms depend upon photosynthesis for growth. Turbidity limits the amount of light penetration in the water. As long as a healthy balance exists between plankton production and the higher forms of life which feed upon them, there will be no bad odors or aesthetically displeasing scums or growths appearing in the water. This balance is a fairly delicate situation to maintain, however, and is easily disturbed.

Domestic sewage from man and his domestic animals frequently pollute the streams and reservoirs. Agriculture practices can contribute excess commercial fertilizers to the water in the form of nitrogen, phosphorous, and potassium. Excess nutrients cause significant changes in the aquatic environment, which can upset the desired species balance. The abundance and species composition of planktonic, bacteria, benthic, and fish populations change with time and at a rate proportional to the level of fertility. Fertilization usually produces undesirable conditions in the reservoir. Chemical enrichment can be desirable, however, in certain situations. For example, efforts to increase production of fish ponds through fertilization have been quite successful in many parts of the world.[4 5] Greater yields are achieved by fertilizing ponds with sewage effluent. However, because of the species changes that typically accompany enrichment in the natural settling of a reservoir, these increases are a major benefit only in areas where the value of fish is determined by weight rather than species diversity.

Biological Resources

Many of the impacts on the biological resources in the project area have already been identified, particularly as they relate to the aquatic environment. The terrestrial species will also be affected. The numbers and composition of species will change. Most faunal species are mobile and will move out and away from the construction activity and away from the water as the reservoir is filled. Animals that cannot find a suitable habitat will die. Others will die because of overpopulation and competition for a specific habitat, but few species will be eliminated. There are exceptions; for example, if there are species in the project area that were already endangered or threatened[35] by extinction before construction, these species may suffer seriously from the impacts of the project and they may be annihilated.

Frequently, there is an increase in the numbers of arthropod disease vectors and nuisance insects. On occasions reservoirs and streams in the United States have been indirectly involved in the spread of diseases, for example, encephalitis by mosquitoes, but never identifiable as the major cause of disease. We are most fortunate in this regard. We have, on the other hand, produced copious numbers of nuisance insects, and at many of our Corps of Engineers projects an insect control program is essential. One of the techniques to control mosquitoes is to fluctuate the level of the pool, causing shallow areas to dry up thus killing the larvae. The natural fluctuations at hydroelectric facilities are helpful, but they can be made more effective if done as part of a deliberate, coordinated plan of operation, based on the known breeding sites and habits of the mosquitoes.

On the beneficial side of the ledger, one usually finds that more fish are produced per pound per acre in the reservoir than existed in the stream before the project was constructed. The pool areas of the project, if properly managed, will result in an increase in the waterfowl population. There will be more opportunities for fishing with larger catches. With proper management of terrestrial areas contiguous to the reservoir, the number of game birds (waterfowl and upland species) will increase and provide greater opportunity for controlled hunting, sightseeing, bird watching, nature trails, and other recreational experiences.

Aquatic weeds cause serious problems for public health, fisheries, irrigation, transportation, and hydropower generation in many tropical regions of the world. In warmer climates of the United States, we find that the shallow waters of reservoirs and waterways provide ideal situations for the growth of aquatic weeds.

It was the rapid infestation of water hyacinths (*Eichhornia crassipes*) in the South Atlantic and Gulf Coastal states, with adverse affects on navigation and reservoir management, that led to the adoption by Congress of the River and Harbor Act, approved in 1899, and subsequent authorizations for projects aimed at removal of this plant species from the waters of these states.

Another introduced aquatic plant which rapidly infested the waters of the Gulf and South Atlantic states is alligator weed (*Alternanthera philoxeroides*). In 1958, Congress authorized a pilot project for Aquatic Plant Control (Public Law 85-500) for the removal of water hyacinths, alligator weed, and other obnoxious aquatic growths from navigable waterways, tributary streams, connecting channels, lakes, and other allied waters.

The rapid invasion of Eurasian water milfoil (*Myriophyllum spicatum*) presented a third aquatic weed species to become a major economic problem in the areas already plagued with a great diversity of aquatic pest plants. In the Chesapeake Bay the plant has spread over 200,000 acres in the past 10 years. Currently, water milfoil is established in the northeast in Massachusetts, Vermont, New York, Pennsylvania, New Jersey, and Delaware. Water milfoil has also been reported in Ohio, Indiana, Illinois and Wisconsin.[27]

The expanded project for aquatic plant control of these different species was initiated July 1, 1967. Section 302 of the River and Harbor Act of 1965 (Public Law 89-298) authorized a continuing program based upon the information presented to the 89th Congress on the pilot project as contained in House Document 251,[40] and testimony presented by the Chief of Engineers (containing data from the U.S. Department of Agriculture, U.S. Public Health Service, and the U.S. Fish and Wildlife Service) to the Subcommittee on Flood Control, Rivers and Harbors of the Committee on Public Works, U.S. Senate, in 1958 and 1962. The River and Harbor Acts were further amended to specifically include Eurasian water milfoil. Provisions for research and planning for the program were included, costs to be borne fully by the federal government, and provisions for control operations were made by which local interests must agree to hold and save the United States free of claims that may occur from control operations, and participate to the extent of 30% of the cost of such operations. Funds for control operations are allocated by the Chief of Engineers on a priority basis, based upon the urgency and need of each area and the availability of local funds.

During the early phases of development of the Corps of Engineers aquatic weed program, many different mechanical devices were designed to cope with the dense infestations of water hyacinths.[38] Early control operations were based almost exclusively on mechanical procedures such as gang-saw boats to open up paths for navigation, physical removal by derrick and grapple, the use of hand labor to cut blocks of matted water hyacinths with 6-ft timber saws, and the use of special barges equipped with conveyor belts for picking up the plants and depositing them on the shore, as well as crusher boats which crushed the plants into a pulp and then deposited the pulp directly into the water.

In these early projects, cutter or destroyer boats were used to open channels clogged with aquatic growths. Types of mechanical control equipment which have been field-evaluated are amphibious tractors, trailers, rolling dollies, wood chippers, an air cushion ground effect machine, and remote-controlled equipment. Mechanical methods for control operations spread obnoxious aquatic plants by fragmentation. Research and field operations have shown that chemical control through the use of 2,4-D (2,4-dichlorophenoxyacetic acid) is the most effective method for control of water hyacinth.

Utility, speed, and mobility are essential requirements for efficient and economical operations. The new program includes a wide variety of equipment types ranging from barge mounted saw boats and spray equipment to mobile air-boats which offer efficient maneuverability. Other types of equipment utilized in control operations range from airplanes, helicopters, inboard and outboard motor boats, trailers, barges, and quarterboats to the use of backpack sprayers.

One relatively new technique which offers real potential for planning, is the use of remote-sensing devices from aircraft for aquatic plant surveys of waterways and reservoirs. Infrared color aerial photography offers such promise because of the economical coverage which can be attained from large areas, and the ability to synoptically view areas in a short time span, during which aquatic plant vegetation usually does not appreciably change. The value of precise ground control in aquatic plant investigations is recognized. The philosophy behind the use of remote sensors is that synoptic small-scale views of aquatic plant problems are of real value in properly assessing the problems of waterways and reservoirs as a whole and reduce the amount of expensive and time-consuming ground control required.

Cultural Resources

Although extensive, the natural environment in the United States and the resources it contains are finite. When our nation was young, the demands of the American people appeared negligible in comparison with the quantities of resources available for them to use. But our population, once small, is now large and is still growing. At the same time our material standard of living is steadily rising. We live in a period of ever-increasing demands for natural resources on one hand, and of ever-diminishing supplies on the other. Clearly, there is a limit to the burden our natural environment can bear, and we must conserve and use our resources wisely.

This phenomenon is in no way restricted to the United States. As the population of the world increases, the stresses on the natural environment of our planet become more critical to man's survival. Recently many people have come to realize that growing demands for resource consumption pose serious threats to their environment; that man's environment is composed of interdependent systems both natural and man-made; and that abuse of one system jeopardizes the quality of the others and ultimately the survival of all.

The National Environmental Policy Act of 1969 (NEPA),[42] passed by the United States Congress, is a dramatic example of this new awareness. I would like to quote a portion of the environmental policy statement from this law to: ". . . encourage productive and enjoyable harmony between man and his environment; to promote efforts which will prevent or eliminate damage to the environment and biosphere and stimulate the health and welfare of man; to enrich the understanding of the ecological systems and natural resources important to the Nation."

The Corps of Engineers[33] has translated this

national policy into four general environmental objectives:

1. To preserve unique and important ecological, aesthetic, and cultural values of our national heritage.

2. To conserve and use wisely the natural resources of our nation for the benefit of present and future generations.

3. To restore, maintain, and enhance the natural and man-made environment in terms of its productivity, variety, spaciousness, beauty, and other measures of quality.

4. To create new opportunities for the American people to use and enjoy their environment.

The feature of NEPA with which the U.S. public is most familiar is the requirement for the preparation of an assessment of the environmental effects resulting from construction of public works projects. This is displayed in the form of an Environmental Impact Statement (EIS). The EIS requires the planning and construction agency[30] to consider and evaluate all environmental aspects of a proposed action. They must also make all the factors considered a matter of public record. They conduct "give and take" sessions in the form of public meetings and consult with state, federal and private organizations on the merits of the proposal. After competitive views are discussed and decisions modified as appropriate, the entire record is placed in the hands of the Council on Environmental Quality (CEQ) which advises the President of the United States on the advantages and disadvantages of the proposed project. The Congress of the United States also receives this information to be used in their decision-making.

What has been the impact of this law on hydropower projects? Most of our projects have been modified in one way or another: dams have been resited, elevation of pools adjusted; fish passage structures designed and constructed; dams built with multilevel water outlet structures; archeological and historical sites located, preserved, or recovered; cemeteries, roads, utilities, homes, and facilities rebuilt or relocated; people relocated and assisted in every way possible to make the move as easy on the individuals as the situation would permit; fish and wildlife losses mitigated, etc., just to name a few of the kinds of modifications the Corps of Engineers has been involved with at hydroelectric facilities.

There have been some indirect benefits accrued as a result of hydroelectric projects. Interest in archeology and paleontology has been stimulated and funds provided for further study. There has been an improvement and upgrading of roads and utilities. Land values have increased with subsequent increases in tax revenues. Trade and services activities have been stimulated in the area of the project, producing a higher standard of living for the people.

REORDERING OF PRIORITIES

The Corps of Engineers and other federal agencies have shown great progress during the period since NEPA, but there are still many improvements which must be made. The American public[43],[47] has become increasingly concerned about the preservation and enhancement of natural values. Great emphasis has been placed upon scenic beauty and protection of fish and wildlife, and at every hand development agencies are enjoined from any action which might adversely affect them. More and more reasons are advanced for why projects should not be built. To an ever-increasing degree engineers are finding themselves in the middle of a conflict of demands — demands on the one hand that our nation develop its resources so that they can be used, and demands on the other hand that these same resources be preserved in their natural state.

These are worthy objectives, but we must be concerned that none of the teach-in programs seem to include any discussion of how to achieve these objectives and at the same time meet the demands imposed by a growing population and a rising standard of living. The present situation is a complex one. Indeed these problems will not be solved by such simplistic proposals as stopping population growth, lowering the standard of living, or stopping the construction of any potential source of pollution, be it solid, gaseous, thermal, noise, or esthetic. Stopping construction does not solve our problems, it merely shifts them to other areas. In fact, it reactivates the same demands that caused construction to be undertaken. The consequences of such extreme measures to protect the environment may well be worse than the ills they seek to cure; nevertheless, we cannot afford to dismiss them out of hand on that basis.

ECOTACTICS

The new word ecotactics has been used recently. It is a hybrid word meaning the tactics to be employed by political activists who are engaged in frustrating some development project by appealing to the public, and ultimately to the courts, contending that its construction will adversely affect the ecology. Ecotactics is rapidly emerging as an effective system, but primarily in a negative way. Those whose business required them to think positively (and engineers are usually in that category) can learn a lot from ecotactics. But if all they learn is better ways of projecting their image, they will have missed the point of the present controversy. What needs to be learned, above all else, is to satisfy the just complaints which have caused it.

The time is past when they can remain passive. Engineers must become increasingly involved in discussing their work not only to justify it, but to indicate to the public why they believe it is in the best public interest. They need not be advocates, but neither do they need to sit by while measures are adopted which are not in the best interests of our nation and its people. They must enter into a dialogue with the environmental activists, and must take great pains not be become sarcastic or to talk down to them. These are intelligent people with whom they are dealing and the great majority of them are honestly seeking the truth. In this dialogue, however, they must themselves be seeking the truth and present their case with a view to convincing them, telling them the truth in terms that they will understand. On the other hand, if they attempt to ridicule those who are raising questions or if they adopt the position that these are technical matters understood only by technicians, they will only destroy the basis for public acceptance of their proposals. And in the final analysis public acceptance is the goal which both sides of the controversy seek.

In seeking that acceptance for work, they must go beyond merely telling the truth when questioned by outside sources. The engineering profession must begin to question itself more intensively and must be just as truthful in its answers internally as to outsiders. Engineers must consider all the consequences of their work and look for alternatives which alleviate the undesirable features of these consequences. In many cases, these alternatives may be nonstructural. One obvious example would be reducing flood damage through flood plain management, an option which can frequently be combined in conjunction with hydroelectric projects. This is a very difficult concept for engineers to grasp. Many will say, "I didn't study structural engineering for four years in college just to adopt a nonstructural alternative." Yet the fact remains that it is their duty as responsible engineers to consider every reasonable alternative and to adopt the best one — in terms of balancing satisfaction of needs with protecting the national environment — whatever it might be.

Actually, when you examine the record closely, you will find that the Corps of Engineers has not been remiss in protecting the environment. On the contrary, they and other federal agencies have achieved a fine record of success in this regard.

The record clearly shows that for many years the Corps has been concerned with the matter of environmental values and is fully capable of serving as a leader in this area, both today and in the future. But all developers must recognize that the definition of and knowledge about environmental values has changed drastically over time, and they must act accordingly. These values have evolved from consideration for wildlife and outdoor recreation, through concerns for natural beauty, to the current emphasis on ecology and the fundamental impacts, positive or negative, of engineering works on natural systems.

Yes, they propose to take care of an environmental problem, caused by engineering, through more engineering works. The environmentalists are likely to ask what the effect of these new engineering works will be. It would be difficult indeed to provide them with completely satisfactory answers based on the types of investigations which have been accomplished to date. Everyone must make more and better use of the growing body of knowledge about the interrelationships between components of natural systems, and to weigh more carefully in the scales of decision the environmental or sociological costs associated with the construction of public works projects.

The Water Resources Council is attempting to develop an evaluation system[46] which will enable federal agencies to compare environmental or ecological values with other project benefits and costs. Although this system has not been refined to the point where everyone is using it, in essence it will involve a determination of the economic opportunities foregone if the project or some of its features are not built in order to protect the environment. Everything the developer does or fails to do involves a cost to somebody. Preserving the environment, even by doing nothing, can be a very costly business if, for example, by doing nothing we continue to incur flood damages or deny to the people who need them the benefits of an adequate water supply, electricity, or cheap transportation.

We must be concerned that our country's future may be threatened today by overenthusiastic preservationists, just as it most certainly is threatened by overzealous developers. The basic problem that confronts those who construct is to reconcile these diametrically opposed views. We must somehow accommodate both of them. We must make adequate provision for the development of our natural resources, hydropower being an excellent example, to support the expanding economy required by a growing population and a rising standard of living. At the same time, we must preserve and enhance those features of our environment which make life in America a thoroughly enjoyable experience. We cannot permit ourselves to be captured by either of these approaches to our country's problems. Without moderation the one would exhaust our natural wealth, while the other would stop the industrial progress which has made our nation great. We could retreat behind the wall by claiming that we need guidance from the administration, from the Congress or from the people, but if we do this what our country needs will probably not be built. Our engineers will point the way, but before they can do this as a profession, they must give serious thought to how they can proceed with work in such a way as to satisfy all the demands that are placed upon them.

Chapter 5
ENVIRONMENTAL PROBLEMS OF LARGE DAMS*

INTRODUCTION

A complex set of environmental factors, including resettlement of people, health hazards, sedimentation, climatic changes, seismic effects, and archeological losses are involved in river basin development. These factors must be evaluated as an integrated total before a satisfactory project can be developed.

The resolution of these problems depends to a large extent on careful planning to avoid situations where these problems are most acute, and by assembling the kinds of technology that are important to cope with those problems which cannot be avoided.

It is also necessary to establish some form of continuing research, in which technically qualified personnel are on site to initiate investigative studies on problems as they become apparent, during the construction and operations phases of development.

From a number of indications it is quite apparent that we as a society have relatively recently begun to ask more sophisticated questions about adverse environmental effects. This has come about, for the most part, because of additional scientific knowledge and more precise investigative tools. By virtue of this increased concern, the realization has come that there may well be factors contributing to the health and/or well-being of man not considered before. Certainly this is a commendable development, but a major dilemma accrues from the fact that there is no real end point to the questioning process.

It should be obvious that decisions in this area will always have to be made without complete information. The goal we seek is a system of establishing values sufficiently flexible to allow us to set tolerance levels that delineate approximate hazards. Because it is the nature of science that experimental results are always subject to further confirmation and refinement, this review must be the basis of an adaptable system manageable within a program of regulation.

We have a further limit, that the more sophisticated investigations become, the more difficult they are to accomplish. There is also a related question concerning the distribution of costs. As more and more background research is required to answer a particular question, additional public involvement is required to determine the environmental effect. The business community, as such, cannot fully undertake the total expense, and more public support is required.

ENVIRONMENTAL FACTORS

The emphasis in this chapter is on the environmental problems caused by man-made impoundments, rather than on the economic benefits derived from them. This is not meant to imply that cost-benefit analysis is not important, nor is it in any way meant to dispute the very real gains which man-made impoundments have achieved under certain circumstances. Rather, this emphasis on environmental problems is the analysis intended in this chapter.[53,60,70]

Preliminary surveys of the main environmental factors involved, and of the kinds of changes which are likely to follow a major inundation of an area, are

1. Flooding will have a substantial effect on the water table in an area surrounding the new lake basin.

2. Many large impoundments create their own climate or microclimate system with the subsequent influence on rainfall, vegetation, and agricultural production.

3. Deoxygenation of the water often follows inundation because of the rotting of drowned vegetation, which in turn often causes an explosive growth in phytoplankton, floating vegetation, and different species of fish.

4. The terrestrial and aquatic flora and fauna will also undergo changes, which will along with other changes tend to increase the variety of human activity and industry.

In presenting this far from comprehensive list, we note that more research of a multidisciplinary kind, working up the food chain from the inorganic and physical factors, is needed, including research on plants, animals, and man himself. This

---

*Abridged and updated from unpublished report "The Consequences on the Environment of Building Dams," U.S. Army Corps of Engineers, Washington, D.C., 1973.

kind of research is especially needed for impoundments which are planned for tropic and subtropical areas, for the tremendous increase in
biological productivity following an inundation
and its return to a new steady level after a period
of years is not well understood. Oxygen depletion,
combined with hydrogen sulfide, increases plankton growth extensively and bottom dwelling fish
may die. There is an increase of mosquito and
larvae vectors, water hyacinth growth, and an
increase of phosphate, nitrate, and organic matter.

Some of the problems which arise from tropical
man-made impoundments have to do with
fisheries, where "maximum sustained yield"
should be the primary goal. A negative effect is
caused by the formation of an obstruction to the
migration of anadromous fishes, coupled with
adverse flooding of upstream spawning grounds.
Changes of temperature and water quality can also
have negative effects. A positive fishery potential
is realized from the increased food supplies and
cover from predators. These changes result in an
immediate removal of many of the pressures
previously limiting the size of the population. Too
much drawdown or siltation can have a serious
negative effect on fishery development, due to the
destruction of spawning grounds.

At some point these pieces must be put
together and looked at as one experience. It is
important that the conflict of the engineer and the
environmentalist be replaced by understanding and
mutual respect for each other's point of view.
Different outlooks and purposes must be honestly
expressed, common goals freely admitted, and
unique abilities recognized. Only when these very
basic philosophical and attitudinal questions have
been taken care of will the solution begin to
appear.

When all is said and done and both sides of the
balance sheet have been added up, are the surroundings and the people for whom the man-made
impoundment was designed, better or worse off?
The question is an extremely difficult one to
answer, but it is one of the first ones that should
be asked when a project is being considered. Will
this effort be a positive and enhancing influence
for the surroundings and people it is meant to
serve? Only early consultation with and inclusion
of a variety of different kinds of professionals into
the initial planning stages of such projects will
make an accurate and realistic response to this
question at all possible.

What kinds of environmental problems and
questions do man-made impoundments cause?
What are the implications of these harmful effects?
Can they be corrected? If so, how? If not, why
not? What is being done to try and eliminate or
minimized them? How successful have these
attempts been? These are some of the questions
we must answer.

RESETTLEMENT OF PEOPLE

Resettlement of people has been one of the
most frustrating and tenacious of all the man-made
impoundment difficulties. When a man-made
impoundment is created, the members of the river
basin population are displaced, crowded, or supplemented by new migrants, and are often stirred
by the political repercussions of enforced relocation. Beyond that population, a much larger area is
affected by the development of new conditions of
life and livelihood. Generally, the larger the lake
the greater and more complex the impacts on the
sociocultural system. One of the hardest aspects of
relocation to accept is the fact that it is always
compulsory. The people in the area have to move
or be moved; they have no choice in the matter.
Such a situation is bound to create resentment and
anti-government attitudes.

Thus the most severe stress as far as the
dislocated people are concerned comes during the
initial period, when they have to make their move
to a new situation, when their old familiar and
familial living, hunting, and burial grounds are
flooded over and forever gone. To observe, let
alone be a part of this phenomena, is traumatic.
There is some evidence that in spite of improved
medical facilities, morbidity and mortality rates go
up during the transition period. Higher population
densities, new diseases, and psychological and
sociocultural stress are probably all contributing
factors in this development. In new surroundings
and with new people, old customs, habits, and
rituals (cultural patterns) may seem out of place
and be discontinued, causing a kind of stress which
is unquantifiable. This initial stress period usually
terminates when the relocatees begin to adapt,
evolve, and cope with their new situation, which
invariably they seem to do in one way or another.

Generally there are four groups of people who
are affected by resettlement: (1) those who must
relocate because their homes and fields will be
partially or totally inundated by the reservoir, (2)

those among whom the relocatees must be settled, (3) lake basin inhabitants who are neither relocatees or hosts, and (4) immigrants who move into the lake basin seeking new opportunities which accompany dam construction and reservoir creation. But the people who are most adversely affected are the relocatees.

This local group of people is part of a sociocultural system which is intricately interrelated with physical and biotic components of the lake basin habitat. A whole way of life is wiped out. Kinship and friendship ties between groups separated by only a river are now separated by a wide body of water. Livelihoods have to be learned all over again; old ways of subsistence farming, fishing, and hunting may not be applicable to the new situation.

Thus, from a human point of view, relocation has been one of the least satisfactory aspects of reservoir projects. Relocation expenses have rarely been less than 25% of the combined cost of power generation, transmission, and dam construction. In some cases the costs of relocation may be so high as to warrant cancellation of the project.

In the cases where adaption does occur, where the original sociocultural system of relocatees is lost, simplification of this complex system results. This situation is dangerous in that it leads to a new demoralization and breakdown. Sociocultural withdrawal often results. People change only as much as they have to in order to continue the realization of relatively fixed cultural goals. People learn to cope primarily by utilizing old behavioral patterns and old premises in new ways. In other words, they change only as much as is necessary to continue doing under new conditions what they previously did and held to be important.

Trying to teach new relocatees new jobs has also been one of the least satisfactory aspects of reservoir projects; this is especially true of the efforts to intensify lake basin agriculture, which has sometimes resulted in over-cultivation and over-grazing and thus massive erosion. Another problem is that the economics of power transmission are such that most of the generating capacity of the turbines is utilized outside the lake basin, of little help to the local people to utilize electricity even if it was available.

Generalizing about the subject of resettlement and sociocultural change is extremely difficult because each peoples' cultural patterns, expectations, and traditions are different from those of the next group. There will, therefore, be various reactions to the same process, not to mention the fact that no two resettlement experiences are the same.

Anthropologists have called these attempts by communities to readjust to a new situation "revitalization movements." These are deliberate, organized, and conscious efforts by some or all of the members of a society to construct for themselves a more satisfying culture. Some feel that in search for new identities, a people can successfully accomplish sweeping changes in cultural behavior in a very short time. Others feel that coping mechanisms utilized are more incremental in nature, whereby the traumatized population attempts to restore its sense of well-being gradually, rather than through a more radical departure from the past.

Whatever the process, the inputs which are required to make resettlement a success are numerous and far-reaching in nature. Systematic background investigations are needed of the rural society to be affected by the creation of a lake, along with an objective appraisal of the extent of the human resources development required by the technical and economic change foreseen. Also required are specifications for the type and timing of supporting facilities needed, the maintenance of social controls, the development of leadership, new local organizations, patterns for administrative relationships with and within the new communities, and provision for succeeding generations.

Past experience has shown that these background studies should go back quite far into the histories of the people being moved. If this step is not taken, all of the provisions which may be provided for them could be completely wrong in terms of what the people are used to. Food will have to be supplied in an extensive grant manner until a resettled or transferred population can reestablish its productive capacity.

Resettlement will also require new buildings for housing, schools, community and health centers, recreation facilities, and public offices, along with storage, market facilities, roads, water supplies, waste removal, and sanitary facilities.

If resettlement is going to have any hope of being a positive experience for those being moved, some way of compensating them for their lost lands and income will have to be devised. Reimbursement will require credit on relatively easy terms, both for the lost land and also in order to help the relocatees start a new vocation. Whether it be fishing or farming, new equipment and supplies will have to be purchased by the individuals trying to make new starts. The lack of viable income generating opportunities in the new communities has quite often been one of the major failings of resettlement efforts. Without an ability to earn an income, less resourceful relocatees often develop attitudes of dependence on government handouts, while the more energetic dislocatees will move out of the newly created towns, leaving "ghost towns" behind.

In order to make all of these inputs function properly, there will have to be extremely good managers on hand to coordinate the various activities. Parts of the lake basin may have to be cleared to allow for efficient and effective fishing techniques, and care will have to be taken to prevent the farmer from developing crops which are already being over produced in the rest of the world. These back-up and managing jobs are critical and often times determine the degree to which a resettlement program will succeed or fail.

One difficulty in the past has been that these kinds of managers and planners have not been integrated early enough into the man-made lakes programs. There has been too little time between the initiation and completion of the dam to carry out and implement the minimal research needed for effective rehabilitation at the time of resettlement. As a result, the move prior to flooding becomes a tension-filled crash program only designed to physically move people into new environments which will not be capable of supporting them on even a subsistence level for years to come.

The incomplete integration of rehabilitation of lake basin inhabitants into the overall project design will in addition often result in resettlement costs three to four times the original estimates.

From this account, resettlement is the most multidisciplinary kind of business, requiring the coordinated effort of a whole gamut of different people with different backgrounds and training. Unfortunately there is no formula for resettlement; it has to be handled case by case. It can probably be said that at least the above various inputs will be needed. Beyond that, care will have to be taken to accurately assess and handle the local situations and cultural nuances of the people being moved. As has been mentioned before, to date resettlement has been a very unsuccessful experience. This failure is probably also due in part to the number of other problems which make their influence felt on the resettlement effort.[56,59,69]

HEALTH HAZARDS

The relationship between health and resettlement is fairly obvious, as are the reasons for disruption of existing health patterns by new impoundments: the substitution of a large, static, seasonally fluctuating stretch of water for a flowing river radically alters the suitability of the area to maintain the intermediate hosts of certain very important diseases. With the building of a dam, and the resettlement of people, large groups of people are forced together in heavy concentrations. A population density such as the area may never have seen before offers opportunities and facilities for a greatly increased transmission of communicable diseases, the whole prevalence of which may be permanently affected.

The following is a list of water-related diseases whose incidence has been observed to increase, at times quite drastically, in different man-made lakes. The description of these diseases will again reveal from another angle the interrelatedness of the problems caused by man-made lakes.

Schistosomiasis (Bilharziasis)

Schistosomiasis (Bilharziasis) and fascioliasis are the two snail-borne infections which might be expected in water development schemes; of the two, schistosomiasis is the greater problem. Schistosomiasis is essentially a water-based disease, being dependent upon an aquatic organism (the snail intermediate host) for a part of its transmission cycle. This serious debilitating disease affects over 200 million of the world's population. The prevalence of the infection is nearly always enhanced by the impoundment of water in man-made lakes and by the irrigation systems frequently associated with them.

Before a lake is formed, the intermediate hosts of *Schistosoma haematobium* and *S. mansoni* (small snails) do not have as ideal a habitat in which to dwell and grow as when the lake or impoundment is formed. This fact is due in part to the succession of plant growth which usually follows. Water hyacinth (*Eichhornia crassipes*) and water ferns (*Salvinia* spp.) will invade the area, eventually coming to rest on the shoreline. Such masses of vegetation, if removed, will be replaced by submerged plants such as coontail (*Cerato-phyllum demersum* L.) which harbor the snail intermediate hosts of schistosomiasis. Thus we see how problems of aquatic plant growth are integrally related to the larger problems of resettlement and health.

In considering this problem it is important to emphasize that it is the definite host, man, who is responsible for the dissemination of schistosomiasis, by contaminating the aquatic environment where he in turn becomes infected, whereas the snail is a passive intermediate host. Therefore, in tackling the epidemiology and control of schistosomiasis, consideration must necessarily be given to the human as well as snail hosts.

The transmission of both urinary and intestinal schistosomiasis occurs in different places for different reasons and with varying intensities, but the above relationship between the plant, snail, and human ecosystems is usually present in some form or other. Generalizing about method of prevention is difficult because of the variety of situations in which schistosomiasis has occurred. In some places clearing of the brush and trees before filling has helped; in other places it has not. At present, chemical and biological attacks on the host snails are the most feasible and profitable method of prevention.

Obviously, the decision to clear trees and other vegetation from areas which will form the bed of a man-made lake, whether reached for navigational, fishery, or public health reasons, calls for the inclusion in the planning of a project at the earliest possible date of qualified health officials and aquatic ecologists who have the ability to determine the best set of preventive alternatives, given the local situation.

Malaria

As a mosquito-borne infection, this disease has caused a great deal of problems for development projects of all kinds, going from the Panama Canal to the Kariba Dam. The lush aquatic vegetation along the shores of many lakes makes ideal breeding grounds for various types of malaria carrying mosquitoes (*Anopheles quadrimaculatus* and *A. gambiae*).

Malaria can be controlled but it takes constant vigilance. If the method of dropping the water level sharply to strand mosquito larvae is used, attention has to be paid to other areas where mosquito breeding might take place. In many countries there is substantial knowledge of the vectors and parasites, with local research and public health workers experienced in the difficult task of control. But dropping of the lake level may adversely affect or completely eradicate fish spawning grounds, which are often found in the shallow areas around the outside perimeter of the lake, another example of how the solution for one environmental problem may be the cause of another.

Again, it is necessary that health officials with a knowledge of local conditions are drawn in at an early stage to participate in the planning of man-made lakes.

Onchoceriosis (River Blindness)

Onchoceriosis (river blindness) is mostly transmitted by *Simulium damnosum* (black fly), whose larvae breed in the rapid flowing sections of streams and rivers. The effects of impoundments are beneficial insofar as they drown out the breeding places for several kilometres above the dam. Although breeding may subsequently take place on the spillways and below the dam, the larvae can be artificially flushed away by opening the sluice gates. The bite of the infected black fly will cause gradual blindness; thus there has been extensive research done on this problem. Onchoceriasis can be controlled through siphonic spillways, submerged pass pipes, and other engineering design features.

Trypanosomiasis

In tsetse fly infested areas, there is an increased danger with the creation of a lake of a recrudescence of trypanosomiasis, or sleeping sickness, since the extensive shorelines of the impoundments will constitute harborages for tsetse flies, wherever the lakes become lined with high vegetation. Control of this disease has been practiced successfully by the prudent use of insecticides and by planned, selective vegetation control.

Filariasis

Filariasis is caused by the mosquito vector *Culex fatigans* and is most severely felt in urban centers where sewage water disposal is inadequate and where polluted water is allowed to lie stagnant. This problem will not arise as a direct effect of a man-made lake, but may be an indirect result following the increased concentration of human population in new villages around the lake. Control of this disease is difficult, but can best be carried out by proper planning for the disposal of lake-side community wastes.

Other Health Problems

In addition to the above diseases which may arise, there are other health difficulties which may develop. If proper food is not supplied for the lake basin community at the right times and in the right amounts, nutrition problems could occur.

Poor diets and low protein levels make people more susceptible to diseases than before, and with high concentrations of people epidemics are more likely. It has been estimated that in the course of a year or two, a population of perhaps 25,000 builds up in the controlled construction zone of a dam, for "in addition to the labor force itself, traders, daughters of joy, beggars and itinerant workers" are attracted to the area.

Experience with the new man-made lakes has shown that the initial financial allocation for public health matters often falls far short of what is needed to adequately deal with health problems arising from such lakes. Health planning along broad lines should be done in the earliest stages or when the dam itself is being planned. Concomitantly, institutional arrangements should be made with existing organizations such as WHO and national and local health authorities.[54,57,61,66]

AQUATIC PLANTS

When a lake or reservoir is deep and covers a large area, plant growth will be minimal. This is because deep water eliminates those plants which have to root on the bottom. Floating plants, which are not hampered by deep water, are subjected to damage by wind and wave action. Where the climate is cold many species of vigorous water weeds are unable to flourish. Such a situation obviously makes the above health considerations not nearly as serious in the more temperate and cooler zones. But in the tropical areas, weed growth and the accompanying health hazards are two serious problems.

Many of the man-made impoundments in the warmer regions of the world have, of necessity, been formed by flooding river valley systems. Thus, many of the lakes are relatively shallow, allowing bottom growing plants to flourish. These impoundments often have complex shapes and hence long margins. An added complication is that sometimes trees are only partially submerged, thereby increasing the complexity of the border of the lake. The recently submerged soil is, moreover, rich in plant nutrients to which is added the breakdown products from the decay of large amounts of vegetation killed by the rising water.

All of these conditions create a situation which is ideally suited to both water plants which root beneath the water and those which float. Many times shortly after and during the formation of a new tropical lake, there is a sharp increase in aquatic plant growth to such an extent that large areas of the lake are taken over with the following results:

1. Navigation by boats becomes difficult or impossible.
2. Hydroelectric installations and harbors can be blocked by large floating mats of weeds.
3. Feeder streams and irrigation outlets can become choked by weeds, which in turn hampers collection and utilization of the water and causes flooding.
4. Such a dense cover can be formed over the surface of the water that fishing can become impossible, or the plants may cause so much deoxygenation of the water that it becomes impossible for the fish to live. Both of these situations will adversely affect the inhabitants of the lake area.
5. Water plants, by their transpiration, may greatly increase water loss from the lake to the atmosphere (called evapotranspiration).
6. Aquatic weeds may substantially reduce the effective storage capacity of the reservoir by occupying large volumes of the water storage region.
7. Extensive weed growth can reduce or eliminate the use of the lake for recreation.

Water weeds provide excellent breeding grounds for many disease carrying insects, snails, and worms. The main plants to guard against are those that float, especially the water hyacinth (*Eichhornia crassipes*). This plant, because of its outstanding ornamental appearance, its powers of vegetative production, its long-lived seeds, and its resistance to attack by pests and diseases, has spread far and wide over the warmer regions of the globe.

There are three main methods of aquatic weed

control: mechanical, chemical, and biological. Ecologically speaking, biological methods are the most desirable and chemical methods the least desirable. However, chemical methods are usually the most effective, though often expensive. Much more research is required on the biological control of aquatic weeds. Mechanical methods of control are expensive and it is difficult to know just how much of this expense can be deferred by harvesting and marketing various aquatic weeds. The question of whether or not aquatic plants can be economically controlled and harvested for use as animal feed is one possibility which needs further observation.

Consultation between biologists, hydrologists, and engineers in the planning stage, and coordinated pre- and post-impoundment surveys and research programs, have shown that it is possible to control weed problems. Mapping of the lake catchment area and identification of potentially dangerous plants will greatly facilitate the successful handling of this problem. Sensible bush clearing will have to be used to reduce the number of weed nurseries. There are many ways in which engineers, through different design features, can incorporate weed control procedures into the dam wall. But again all of this will require coordination between members of what is now becoming a man-made lakes team.[61,62,66]

THE FISHERIES PROBLEM

In the past, reservoirs have been created primarily for irrigation, water storage, flood control, and hydroelectric purposes. Rarely, if ever, has the building of a reservoir been justified solely for the purpose of increasing fish production. The rapidly increasing demands for protein, however, especially in the more affluent societies, now make the fullest use of man-made lakes for fishery development all the more urgent. Above and beyond these reasons is the immense value of fishery development for the physical and mental health of the people and the biological systems of those countries concerned.

It is therefore essential both to learn how to manage reservoir fisheries to give maximum sustained yields of desired species, and to ensure that damming the river does not affect already existing fisheries in the river system, both upstream and downstream of the new lake, and in the estuaries and neighboring sea. Succinctly stated, the goal of reservoir fisheries is usually the production of as much fish, of the most desirable species as possible, and the maximum yield that can be sustained year after year by commercial and/or recreational fishing without damage by overfishing.

After damming a river, the initial biological explosion in some man-made lakes is due both to the spatial expansion of the aquatic environment and to the release of nutrient materials from bottom soils and submerged plant and animal remains. The rapid increase in microbiological populations results in sharply increasing fish levels for the first two or three years of a lake's existence. This peak will then level off and in some cases decrease a little during the period of lake stabilization. There is some dispute over this point. Some feel that reservoir productivity increases rather than decreases with the aging of a reservoir. The first few years of an impoundment are generally characterized by high rates of deoxygenation caused by decomposition of inundated vegetation and/or organic debris. In instances where existing vegetation is not cleared prior to reservoir filling (e.g., the Volta Lake in Ghana) deoxygenation may be extensive and lead to significant reductions in the original riverine fish population.

Problems which may arise have to do with fish feeding, aquatic weeds, fish diseases, and change in species composition. Most species of fish which breed in fast flowing rivers normally swim against the currents when feeding. When such fish are kept in an almost stationary body of water, their feeding capacity is adversely affected, especially if the water is not very productive.

Perhaps the greatest single problem associated with the establishment of a fishery in a new reservoir is the overpopulation of the reservoir with large, rough, aggressive fish. These fish tend to have a high fecundity, are biologically efficient, and are able to live in conditions where other important fish cannot. Possible solutions include the use of selective pesticides, the use of predators and parasites, and the planting of forage fish capable of competing with the problem species.

A further problem involves loss of spawning areas due to both sedimentation and drawdown. Silt accumulation often covers over good spawning areas and increases the opportunity for the loss of eggs to predating fish. Sedimentation also covers many of the benthic organisms that form part of various fish diets. Drawdown has a definite impact on both spawning and general fishing success in reservoirs. Because drawdown tends to minimize the growth of littoral vegetation, it inhibits spawning by reducing the necessary cover for eggs; this lack of vegetation minimizes the habitat of the various warm water species that inhabit the littoral zone around the shoreline.

As mentioned in the last section, aquatic weeds can also cause difficulties for man-made lake fisheries. Dense growths of plants cause excessive shading, deoxygenation, and a consequent reduction in primary production. As with the appearance of new aquatic weeds, so also do certain fish parasites and diseases which are uncommon in fast flowing rivers become established in man-made lakes, decreasing fish production in the process.

Deeper parts of the lake may become uninhabitable for fish, as higher concentrations of toxic hydrogen sulfide build up near the dam.

The negative effects which a dam has on migratory fish are the most serious in terms of fishery development. These effects influence both the adults going upstream and the young migrants going downstream. In some cases a reservoir may inundate a spawning ground for migrant fish. The problems of movement of mature adults and downstream juveniles are much more serious.

For adults the problem created by the lake has to do with the disturbance of the timing and energy budget for their upstream migration. The most common mistake in the design of fish ladders is to assume that the fish will know that a device has been put there for their benefit. The major problem is to get the fish into the devices without delay. If a dam is over 100 ft in height, it is usually not possible to construct facilities that will move the adults upstream past the dam without delaying their migration substantially.

The consequence of this kind of delay in migration is that fish may reach the spawning grounds at the wrong time, and they may have also exhausted their energy reserves and die without spawning. For the young fish going seaward, there are also difficulties. Many times they get lost, delayed, or trapped going from the head of the lake to the dam. Those that do get to the dam will many times die going over the dam or through the turbines, or they will die from abrasion on the spillway or tailrace. To date, although many methods have been experimented with (by-pass

channels, louvre systems, light and sound guides) no completely satisfactory solution to these problems has yet been found.

It is thus quite apparent that where a very large dam or series of dams is to be built, one should not be optimistic about the preservation of migratory fishes. Quite often extreme cases have been observed in which certain species of fish have totally disappeared from man-made lakes after some years.

In order to minimize these adverse effects on fisheries, biologists need to be included in planning teams from an early stage, and to work closely with engineers and others whom we have recommended should be included on this man-made lakes team. Clearly coordination of these many different disciplines will be a problem.[48-51, 54,63,65,67]

SEDIMENTATION

Before a free flowing river is dammed to create a lake, it carries throughout its course a number of different kinds of sediments which are kept in suspension by turbulence. When the river is dammed, however, a reservoir is formed, and the river's velocity and thus the extent of its turbulence is greatly reduced. As a result, a portion of the incoming suspended sediment load will be deposited or "trapped" in the reservoir and lead to the formation of a "delta." Although it is difficult to predict the precise form that a delta will take, a number of empirical methods have been developed to approximate the extent and form of a delta over the life of a reservoir.

The time it takes to completely fill a reservoir with sediment can vary anywhere from several years to several thousand years, depending on the climate, the extent to which lake levels are varied, and so on. Another problem which springs out of reservoir sedimentation has to do with the deposition of sediments in the stream channels just upstream of the lake, called aggradation. As a result of this phenomena, the level of the river in back of the lake has been known to increase by as much as 13 ft, causing extensive flooding, seepage, elevation of the ground water table, swamping or waterlogging of adjacent lands, and the formation of stagnant pools.

Other problems associated with sedimentation involve the formation of vegetative growth on the delta, which causes more evapotranspiration, and a further slowing down of the flow of water, allowing a greater quantity of sediments to settle out. The process of sediment trapping can also influence the water quality of a lake in at least two ways. For one thing, sedimentation generally leads to a reduction in the turbidity of the waters, which while being useful for human activity, nevertheless can lead to increased photosynthetic activity since light penetration is increased, which in turn causes an increase in biological productivity, and a hastening of the process of lake eutrophication.

Secondly, the water quality is affected by sedimentation through the changes in chemical reactions that take place at the sediment-water interface. The deposition of the suspended load on a reservoir bottom creates a readily available pool of minerals and nutrients which can potentially cause water quality deterioration. The most graphic example of how sedimentation can increase the concentrations of harmful substances concerns the transport of pesticides and heavy metals. Sediment transport can be an important means by which these substances are circulated throughout the environment, and infused into the various food chains. The degree to which these events occur, and how they occur, requires further investigation.

The problem of premature filling of a lake basin is also exacerbated by reservoir bank erosion. It is often difficult for vegetation to establish itself on the banks of reservoirs that are subject to large fluctuations in water surface elevation. As a result, the shoreline may remain bare in places, and exposed to erosion from either rainfall or wave action. If a reservoir has long reaches of its area which coincide with the direction of the prevailing wind, then significant shoreline erosion may result.

The magnitude of erosion in reservoirs is often greater in arid regions than in wetter regions. The erosion of the shoreline increases turbidity and introduces harmful nutrients and/or minerals into the reservoir. The obvious solution to this problem is to try to stabilize the lake level, thus allowing vegetation to establish itself along the perimeter of the lake. This solution is, however, difficult to bring about, especially in the tropics where wet and dry seasons are prevalent. Terrestrial biologists and other land conservation experts will be necessary in order to start working toward a solution to this difficult question.[51,52,55,58,62,69]

WATER QUALITY CHANGES

There are a number of reasons why the transformation of a free flowing stream to an impoundment can lead to changes in water quality at the reservoir site. One reason is that the waters have a much increased "detention time," which provides a potential opportunity for slow reactions to come closer to completion than they can in the rapidly moving waters of a river.

Another reason is that thermal stratification creates a number of distinct regions within the water body. In discussing thermal stratification (the formation of density layers in a reservoir resulting from variations in water temperature) it is helpful to use the following terminology: the "epilimnion" refers to the well-mixed upper stratum which may be 30 to 50 ft deep; "thermocline" refers to the layer below the epilimnion, in which temperature decreases rapidly as depth increases; and "hypolimnion" refers to the bottom stratum.

One of the main environmental quality concerns associated with thermal stratification has to do with the distribution of dissolved oxygen throughout the reservoir. Oxygen can enter a reservoir through surface re-aeration and photosynthetic action. Since both of these processes occur in the epilimnion, that layer is well oxygenated. The stratification prevents convective re-aeration of the hypolimnion; oxygen diffusion across the thermocline is a very slow process, and thus the hypolimnion receives a very low influx of dissolved oxygen. The hypolimnion, however, generally has a biological oxygen demand (BOD) resulting from the decay of dead organisms which sink from the epilimnion. In this way the hypolimnion can easily become deoxygenated.

Oxygen depletion in the hypolimnion is often considered the most serious consequence of thermal stratification. As long as a positive dissolved oxygen concentration is maintained throughout the hypolimnion, most of the adverse water quality problems of stratification are avoided. Once the hypolimnion becomes anaerobic, however, an entirely different chemical regime is set up in the reservoir.

Dissolved oxygen is one of the most important water quality parameters in a reservoir. Dissolved oxygen must be present in certain minimum concentrations to preserve fish life in the water. Also, adequate concentrations of dissolved oxygen prevent increases in iron and manganese concentrations. Oxygen is finally important in terms of organic decay. All lakes contain decaying organic matter, but man-made lakes, as we have seen, contain an abundance of decaying material. In the presence of oxygen, decay yields the innocuous substances CO_2 and H_2O. In the absence of oxygen, decay of the same organic matter may release H_2, H_2S, CH_4, and NH_3, all of which are undesirable in water supplies.

Dissolved oxygen problems are most apt to occur in the hypolimnion during periods of thermal stratification. The hypolimnion receives its dissolved oxygen through convection during the fall and spring overturns. During the rest of the year a low rate of re-aeration occurs by oxygen diffusion across the thermocline. As a result of the inundation of vegetation and humus soils, the hypolimnion of an artificial reservoir is likely to have a high chemical oxygen demand (COD) when the reservoir is first filled. Added to this initial COD is the BOD from those organisms which die in the epilimnion and sink to the hypolimnion. Typically, without artificial mixing, the hypolimnion will be anaerobic during the first few years of an artificial reservoir's existence. Once the initial COD is oxidized, the hypolimnion can revert to an aerobic state.[54,57,61,68]

EUTROPHICATION

This subject draws together many of the other topics which we have been discussing, for the water quality of a lake is influenced either directly or indirectly by almost every activity which takes place on, in, or around the lake, whether it be boating, fishing, power generation, agriculture, forestation, or just plain living.

Lake eutrophication is an economic, recreational and aesthetic problem that affects all the lakes in the world. It is the natural process of lake aging, and would progress even if man were not present. Man's pollution, however, can hasten the natural rate of aging and shorten the life expectancy of a body of water. The eutrophication of a lake consists of the gradual progression from one life stage to another based on the degree of nourishment. The extinction of a lake is attributed to enrichment by nutritive materials, biological productivity, decay, and sedimentation.

Normally, with a lake, the youngest stage of its life cycle (the oligotrophic stage) is characterized by low concentrations of plant nutrients and little biological productivity. With a man-made lake, this youngest stage is skipped over; the lake is in a way dropped down in *medias res*. From the time that it is formed, it has fewer years to live than a newly formed natural lake; for in a new man-made lake the concentrations of plant nutrients and biological productivity are relatively high, because of the comparatively large quantities of plants, trees, and other living organic objects which when flooded will die and start to decompose, releasing large quantities of nitrogen and other nutrients in the water. This phase in a lake's existence is called the "mesotrophic" stage. During this time biological activity within the lake increases and organic sediments accumulate on the bottom of the lake.

Because a new man-made lake starts its life at middle age, special care has to be taken to ensure that the process is not speeded up any more. Enrichment and sedimentation are the principle contributors to the aging process. The shore vegetation and higher aquatic plants utilize part of the inflowing nutrients, grow rapidly, and in turn trap sediments. The lake gradually fills in, becoming shallower by the accumulation of plants and sediments on the bottom, and smaller by the invasion of shore vegetation, eventually becoming dry land. The extinction of a lake is, therefore, a result of enrichment, productivity, decay, and sedimentation.

The eutrophic cycle involves the establishment of nutrient concentrations at levels that permit the excessive growth of algae. The nutrients tend to feed and accelerate an algal bloom in the epilimnion as nutrients are circulated in the lake. Algae from the bloom will die and sink to the hypolimnion where they are decomposed by bacteria, releasing the nutrients to again feed the bloom in a cyclical fashion. The decomposing algae often exert a biochemical oxygen demand (BOD) in excess of the available oxygen, thus depleting the dissolved oxygen (DO) of the hypolimnion. The resulting low DO concentrations, combined with the toxic effects of the decomposing algae, may cause massive fish kills.

Human activity of many kinds can and does increase the extent and speed of these processes, a development which is particularly disconcerting with regard to newly created man-made lakes, which start their lives at unusually high levels of nourishment and productivity.

Human Wastes

Human wastes are the most obvious and first sources of difficulty with regard to the nutrient level of the new lake. Raw sewage from both animals and humans pumped daily into a new lake will soon significantly change the nutrient level of the lake, especially if the population density of the area is increasing, as is more than likely.

A strong effort must thus be made to create sewage collection and disposal systems. The resulting health benefits, in addition to increasing the lifetime of the lake, include prevention or reduction of insect-borne diseases.

Agricultural Wastes

Another common source of nutrients is drainage from agricultural lands, which often contains nitrogenous or phosphorous-based fertilizers. Drawdown and foreshore agricultural procedures can perhaps be used to withdraw these nutrients from the lake curtain area. Contour planting and strip cropping will help to hold the soil and run-off water in place, but this advanced farming technique requires careful planning.

Erosion and Forestries

Erosion and forestries have a close cause and effect relationship. If forestries are well managed, erosion and thus water pollution will decrease and lake eutrophication slow down. If they are poorly managed, erosion will ensue.

The manner in which logging is carried out can have important effects on both streamflow and water quality. If adequate regeneration is slow, operations involving large-scale clearing or burning can be followed by increased flood peaks and sediment loads. The extensive use of insecticides and herbicides in forestry, as in agriculture, can have serious repercussions on lake productivity and water quality.

The reduction of soil erosion and the improved infiltration and water retention capacity of forest soils, combined with the eventual provision of commercial wood crops, can both directly and indirectly raise the general level of catchment protection.

Industrial Pollution

Industrial pollution will no doubt increase as new businesses start to crop up. Of particular interest here is the effect of thermal discharges from hydroelectric plants located on or near the new lake. The work done on this problem has resulted in several expensive but unsatisfactory responses.

Sedimentation

Sedimentation and enrichment are the principle contributors to the aging process of a lake. The two factors reinforce each other. Increased enrichment causes increased plant growth. The plants in their turn are partly responsible for trapping sediments.

All streams transport some sediment, and there is a natural tendency for this sediment to be deposited when streams enter a dry basin or body of water impounded behind a dam. Depletion of storage capacity is but one of the effects of reservoir sedimentation. The stream channel is likely to aggrade for some distance above the reservoir. Flooding will occur more frequently, and drainage of flood plains will be impeded because of reservoir sedimentation.

Downstream effects of reduced sedimentation include channel degradation and streambank erosion. The influence of sedimentation on lake nutrient and light penetration levels is still largely speculative.

The relationship between the amount of sediment remaining in a reservoir and the amount delivered to it is termed reservoir trap efficiency. Many variables seem to determine the degree of trap efficiency: among them are properties of influent sediment, type and location of spillways, dissolved solids in the water, and many other unknowns.

Most of the sediment delivered to reservoirs is the product of soil erosion in the watershed area draining into them. It comes from farmlands, rangelands, woodlands, roads and highways, urban developments, and construction sites.

It is much easier and cheaper to try and keep sediment out of a reservoir than to cope with it after it gets into the reservoir. Once it is there, however, the discharge of sediment-laden waters through the outlet works of dams is sometimes effective. Dredging is another method for controlling sedimentation, but is usually too costly to make it worthwhile.[50,52,62-64]

CLIMATIC CHANGES

Depending on a multiplicity of factors,

including moisture content, temperature, and movement of air masses, along with regional topography, compass orientation, and the size of the reservoir, local microclimate and even gross weather may be changed by man-made lakes. The eventual chain of events among animals or plants may have even stronger secondary effects on human activities.

With the creation of a lake the vegetative land surface is replaced by a water surface. Theoretically, this means that more water would be available for evaporation, thus increasing atmospheric moisture. The interaction of a lake and the atmosphere occurs mainly through the exchange of mass, heat, and momentum. In addition to increased rainfall, some cooling effects have been recorded after the creation of some man-made lakes. Increased snowfall has also been attributed to the building of particular dams.

There is further evidence that lakes tend to suppress thunderstorm activity in the early summer and that their influence on winds is a factor to be considered in planning the control of air pollution. These considerations on air pollution are also important as they relate to air-water interchange and the qualitative aspect of precipitation washouts, which pollute man-made and natural lakes.[52,54,55,58,64,65]

SEISMIC EFFECTS

The filling of a man-made impoundment has sometimes imposed new stresses on the earth's crust, which in turn have generated seismic movements and in some cases earthquakes of a severity (Richter scale of 6.4) which have caused human losses. These vary in magnitude and time in accordance with a number of factors. Water height of 100 m or more in a reservoir constitutes a factor which may be of major seismic importance in combination with geological formation and structure.

Generally, the seismic movements build up slowly to a peak several years after the reservoir begins to fill and then gradually decline. A man-made lake is more likely to have seismic difficulties than a natural lake, because of the comparatively short time it takes a man-made lake to fill. Greatly increased pressure is placed on the earth's crust over a short period of time. Moreover, the saturation of sedimentary formations by seepage from the reservoir may not only cause major losses of water but additional seismic effects.

It has been argued that the mass of stored water in artificial impoundments has led to a number of "artificial earthquakes" in various locations throughout the world. Examination of maps showing regions that can be considered to be effectively free from tremors has made it quite clear that the construction of lake reservoirs has produced earthquakes in areas that have been previously listed as free from tremors. An example is provided by Lake Mead on the Colorado River. During the 15 years preceding the impoundment, the area experienced no earth tremors. After the reservoir was filled, several thousand movements were recorded. Geologists who have studied the situation believe the tremors were caused by rock tensions resulting from increases in the water load.

Unless account is taken of these stresses by prior study and observations, both the dam and nearby areas may be subjected to unexpected damage. Anticipation of such effects becomes an essential part of feasibility investigations. Preliminary studies should include detailed examination of ecological and geomorphological conditions of the reservoir areas.[51,64,66,69,71]

ARCHAEOLOGICAL LOSSES

Archaeology may be narrowly defined as the recovery from the earth of the material remains of man and his works. From the study of those remains, archaeologists try to reconstruct something of the pattern of the past. It has long been known that river basins have been the centers of cultures all over the world. Civilizations have down through the ages been drawn to the many benefits which rivers can provide. The past record of these different peoples is thus most easily found in the various river basins of the world.

Today other men also look at those rivers as potential sources for man-made lake creation. All too often in the past these reservoirs have been created before archaeologists have had sufficient opportunity to survey the area to be flooded, and to excavate the important sites. Similarly, people were moved before historians and social scientists could record their traditions and study their customs and interrelationships with the river basin habitat. In the future, as mankind become increasingly concerned about and interested in the past, he cannot afford to destroy his history without adequate records first being obtained.

For the planner, this concern for archaeological and historical information has both a pragmatic and a theoretical rationale. It is practical in that information about the past is a part of the local heritage and character, which nationalities may wish to incorporate into their schools and their national culture. This information is also a crucial input into the successful resettlement program. Only by knowing people's past expectations can you expect to know their present reactions.

This chapter is intended to give some indication of the complexity and scope of the problems which man-made lakes can create. It should be noted that this account is not a complete one. For example, physical impacts downstream, changes in terrestrial ecosystems, and many other problems have not been touched upon. It is merely an attempt to try to familiarize those concerned with the general problem areas and what will be required to deal with them. At a minimum the following disciplines will have to be integrated into the planning of a man-made lake project at the earliest possible date: sociologists, hydrologists, health experts, meteorologists, government officials, archaeologists, historians, geologists, forestry and agricultural experts, engineers, and many other kinds of people. The coordination of such a team will also present serious difficulties in

that the work of each of the various disciplines will have to be carefully integrated into the project and the plan as a whole.

No one aspect of the program can be considered by itself, as different facets of man-made aquatic ecosystems are closely related and dependent on each other. There is a complex and tightly interrelated system of feedback relationships. In the face of such an array of problems, it is easy to become pessimistic and negative about the prospects of man-made impoundments.

The problems they cause are so extensive in time and space that it becomes difficult to grasp the entire significance and effect which they are actually having. One never really knows how one is progressing in terms of solving or answering the question of whether man-made impoundments are doing more harm than good; for once this initial dramatic and extensive incursion is made into a riverine ecosystem, a set of complex and interrelated counterreactions are set in motion, which will in turn require additional corrective measures.[50,51,54,61,67,71]

A world register of man-made impoundments is given in Table 5.1, to indicate the scope of operations that are involved in a consideration of the total area and the particular countries in the world that are involved in this effort. This is most important because, as other energy sources are used up, hydropower becomes more and more significant.

TABLE 5.1

The World's Reservoirs with the Surface Area of 1000 km² or More

Name of lake	In operation since	Country, river	Area (km²)
1. Volta (Akosombo)	1965	Ghana, Volta	8730
2. V. I. Lenin (Kuybyshev)	1955	U.S.S.R., Volga	6500
3. Churchill	Under construction	Canada (Labrador), Hamilton	6200
4. Bukhtarma (Lake Zaysan with 1900 km is included in this reservoir)	1960	U.S.S.R., Irtysh	5500
5. Bratsk	1961	U.S.S.R., Angara	5426
6. Nishne Kamskaya	Under construction	U.S.S.R., Kama	5400
7. Kariba	Filling started 1958 Full capacity 1963	Rhodesia/Zambia, Zambesi	5180
8. Nasser (Sadd-el-Ali)	1968	U.A.R., Nile	5250
9. Rybinsk	1941	U.S.S.R., Volga	4550

TABLE 5.1 (continued)

The World's Reservoirs with the Surface Area of 1000 km² or More

Name of lake	In operation since	Country, river	Area (km²)
10. Kamenskoye	Completed	U.S.S.R., Ob	4500
11. Ceder Lake Reservoir (Grand Rapids)	1964	Canada (Manitoba), Saskatchewan	4100
12. Cheboksary	Under construction	U.S.S.R., Volga	3780
13. 22.8 Congress (Volgograd)	1958	U.S.S.R., Volga	3160
14. Tsimlyansk	1952	U.S.S.R., Don	2700
15. Caboro Bassa	Under construction	Macambique, Zambesi	2700
16. Sanmen	1962	China, Hwangho	2350
17. Kremenchug	1961	U.S.S.R., Dnepr	2500
18. Kakhovka	1955	U.S.S.R., Dnepr	2155
19. Krasnoyarsk	1966	U.S.S.R., Yenisey	2130
20. Conservation Area No. 3A (Everglades)	1963	U.S.A. (Florida)	2038
21. Wilyuiskaya	Under construction	U.S.S.R., Wilyui	2010
22. Saratov	1965	U.S.S.R., Volga	1950
23. Manicouagan 5	1968	Canada (Quebec), Manicouagan	1942
24. W.A.C. Bennett (formerly Portage Mountain)	1968	Canada (British Columbia), Peace	1761
25. Kama	1954	U.S.S.R., Kama	1720
26. Furnas	1965	Brazil, Rio Grande	1606
27. Sounda	Under construction	Congo (Brazaville), Kouilou	1600
28. Gor'ky	1955	U.S.S.R., Volga	1570
29. Afobaka (dam) Brokopondo (lake)	1967	Surinam, Suriname	1560
30. Oahe	1965	U.S.A., Missouri	1520
31. Kossov	Under construction	Ivory Coast, Bandama	1500
32. Garrison	1960	U.S.A., Missouri	1488
33. Seul Reservoir	?	Canada (Ontario), English	1396
34. Gouin Reservoir (La Loutre)	1917	Canada (Quebec), St. Maurice	1295
35. Lake Manouane	1961	Canada (Quebec), Manouan	1290
36. Kainji	1969	Nigeria, Niger	1243
37. Rincon des Bonete	1946	Uruguay, Rio Negro	1140
38. Tres Marias	1961	Brazil (Minas Gerais)	1130
39. Votkinsk	1962	U.S.S.R., Kama	1120
40. Novosibirsk	1957	U.S.S.R., Ob	1070
41. Kelsey	1960	Canada (Manitoba), Nelson	1012

Adapted from Lagler, K. F., in Berkowitz, D. A. and Squires, A. M., Eds., *Power Generation and Environmental Change*, M.I.T. Press, Cambridge, Mass., 1969, 79.

REFERENCES – PART I

1. Anon., *Cleaning our Environment; The Chemical Basis for Action,* American Chemical Society, Washington, D.C., 1969.
2. Cain, S. A., *Biotope and Habitat of Future Environments,* Natural History Press, Garden City, N.Y., 1966.
3. Ewel, K. C., Braat, L., and Stevens, M. L., Use of models for evaluating aquatic weed control strategies, *Hyacinth Control J.,* 13, 34, 1975.
4. Garman, W. H., Agriculture's place in the environment; considerations for decision making, *J. Environ. Qual.,* 2, 327, 1973.
5. Hestand, R. S. and Carter, C. C., The effects of a winter drawdown on aquatic vegetation in a shallow water reservoir, *Hyacinth Control J.,* 12, 9, 1974.
6. Lindeman, L., The trophic-dynamic aspects of ecology, *Ecology,* 23, 399, 1942.
7. Loomis, W. E., Photosynthesis, the major enzymatic process, *Trans. Am. Assoc. Cereal Chem.,* 9, 48, 1951.
8. Anon., *Pest Control Strategies for the Future,* National Research Council, National Academy of Sciences, Washington, D.C., 1972.
9. Ovington, J. O., Quantitive ecology and woodland ecosystem concept, *Adv. Ecol. Res.,* 1, 103, 1962.
10. Odum, H. T., *Environment, Power and Society,* Interscience, New York, 1971.
11. Rabb, R. L. and Guthrie F. E., Eds., *Concepts of Pest Management,* North Carolina State University, Raleigh, N.C., 1970.
12. Spencer, D. A., *The National Pesticide Monitoring Program,* National Agricultural Chemicals Association, Washington, D.C., 1974.
13. Swingle, H. S., Control of pondweeds by use of herbivorous fishes, *Proc. South. Weed Conf.,* 10, 11, 1957.
14. Stevenson, J. H., Observation on grass carp in Arkansas, *Prog. Fish Cult.,* 27, 203, 1965.
15. Anon., National survey of needs for hatchery fish, Resource Publication 63, Bureau of Sport Fisheries and Wildlife, U.S. Department of Interior, Washington, D.C., 1968.
16. Cummings, J. G., Use, regulation and registration of chemicals used in fish culture and management, *Symp. Registration and Clearance of Chemicals for Fish Culture and Fishery Management, 99th Ann. Meet. Am. Fisheries Soc.,* New Orleans, 1969.
17. Walker, C. R., Toxicological effects of herbicides on the fish environment, *Water Sewage Works,* 3(3), 113, and 3(4), 173, 1964.
18. Engineer Regulation 1105-2-507, Preparation and Coordination of Environmental Statements, Office of the Chief of Engineers, Washington, D.C., 1973.
19. Engineer Regulation 1105-2-105, Guidelines for Assessment of Economic, Social and Environmental Effects of Civil Works Projects, Office of the Chief of Engineers, Washington, D.C., 1972.
20. Engineer Regulation 1165-2-500, Environmental Guidelines for Civil Works Program of the Corps of Engineers, Office of the Chief of Engineers, Washington, D.C., 1970.
21. Ash, C. G., Three-year evolution, Corps of Engineers, *Water Spectr.,* 5(4), 31, 1973.
22. Ash, C. G., Corps Problems in Preparing EIS, Proc. Fortieth North American Wildlife and Natural Resources Conference, March 16 to 19, Pittsburgh, Pennsylvania, 1975.
23. Bell, M. C., Fisheries Handbook of Engineering Requirements and Biological Criteria, Fisheries Engineering Research Program, Corps of Engineers, North Pacific Division, Portland, Oregon, 1973.
24. Coryell, R. B. et al., Geothermal Energy Program, National Science Foundation Report, Stock No. 039-000-00283-0, U.S. Government Printing Office, 1976.
25. Anon., Environmental Quality – 1975, Sixth Annual Report, Council on Environmental Quality, Washington, D.C., 1975.
26. Anon., Environmental Quality – 1976, Seventh Annual Report, Council on Environmental Quality, Washington, D.C., 1976.
27. Crowell, T. E., Steenis, J. H., and Sincock, J. L., Recent observations of Eurasian watermilfoil in Currituck Sound, North Carolina, and other coastal Southeastern States, Report of North Carolina Wildlife Resources Commission, Raleigh, North Carolina, and Bureau of Sport Fisheries and Wildlife, U.S. Fish and Wildlife Service, Patuxent Wildlife Research Center, Laurel, Maryland, 1967.
28. Anon., Consideration of Esthetic Values in Water Resource Development, Corps of Engineers, ER 1165-2-2, 1967.
29. Anon., Preservation and Enhancement of Fish and Wildlife Resources, Corps of Engineers, ER 1105-2-129, 1973.
30. Anon., Preparation and Coordination of Environmental Statements, Corps of Engineers, ER 1105-2-507 (33 CFR 209.410), 1974.
31. Anon., Planning Processes: Multiobjective Planning Framework, Corps of Engineers, ER 1105-2-200 (33 CFR 290), 1975.
32. Anon., Hydropower – The Role of the Corps, Office of the Chief of Engineers, Corps of Engineers, EP 1162-2-3, 1976.
33. Anon., Water Resources Policies and Authorities: Environmental Policies, Objectives, and Guidelines For Civil Works Programs of the Corps of Engineers. Corps of Engineers, EP 1165-2-501 (41 CFR 210), 1976.
34. Anon., Fishtrap Lake, Kentucky, Informational Data on Section 216 Survey, Corps of Engineers, Huntington District, Huntington, W. Va., 1976.
35. Endangered Species Act, 1973, Public Law 93-205, United States Congress, Washington, D.C. (87 Stat. 884).

36. Federal Water Pollution Control Act, 1972, Public Law 92-500, United States Congress, Washington, D.C. (66 Stat. 816).

37. **Anon.,** Federal Hydroelectric Plants in Operation, Under Construction, and Authorized, January 1, 1976, Annual Tabulation, Federal Power Commission, 1976.

38. **Guscio, F. J., Bartley, T. R., and Beck, A. N.,** Water Resources Problems Generated by Obnoxious Plants, *ASCE J. Waterw. Harbors Coastal Eng. Div.,* 91(4), 47, 1965.

39. **Ash, C. G.,** Hydropower and the Environment, National Science Research Council, Georgetown, Guyana, October 4 to 8, 1976, Guyana University Press, proceedings in preparation.

40. Expanded Project for Aquatic Plant Control, House Document No. 251, 89th Congress, 1st Session, U.S. Government Printing Office, Washington, D.C., 1965.

41. **Leman, B. D.,** Annual Report of the Biological Section of the Engineering Department, Public Utilities District No. 1, Chelau County, Wenatchee, Washington, 1968.

42. National Environmental Policy Act, 1969. Public Law 91-190, United States Congress, Washington, D.C. (83 Stat 852; 42 U.S.C. 4221 *et seq.*).

43. **Ortonlano, L.,** Water plan ranking and public interest, *ASCE J. Water Res. Plan. Manage. Div.,* 102, 35, 1976.

44. **Schoeneman, D. E., Pressey, R. T., and Junge, C. O., Jr.,** Mortalities of Downstream Salmon at McNary Dam, *Trans. Am. Fish. Soc.,* 90, 58, 1961.

45. **Swingle, H. S. et al.,** Reports of Fishculture Investigations: India, Japan, South Vietnam, Paraguay, East Pakistan. Agriculture Experiment Station, Auburn University, Auburn, Alabama, Reports to AID, 1968—1970.

46. **Anon.,** Principles and Standards for Planning Water and Related Land Resources, Water Resources Council, 10 September (38 F.R. 24778 to 24869), 1973.

47. **Wileke, G. E.,** Identifying publics in water resources planning, *ASCE J. Water Res. Plan. Manage. Div.,* 102, 137, 1976.

48. **Anon.,** Reservoir Fisheries and Limnology, Special Publication No. 8, American Fisheries Society, Washington, D.C., 1970.

49. **Anon.,** Reservoir Fisheries Reservoir Symposium, University of Georgia Center for Continuing Education, Athens, Georgia, American Fisheries Society, Washington, D.C., 1967.

50. **Anon.,** Proceedings of the International Symposium on River Ecology and the Impact on Man, American Fisheries Society, Washington, D.C., 1971.

51. International Symposium on Man-Made Lakes, Knoxville, Tennessee, American Geophysical Union, Washington, D.C., 1973.

52. **Anon.,** Lake Erie: Dying but not dead, *Environ. Sci. Technol.,* 1, 212, 1967.

53. **Lagler, K. F.,** Ecological effects of hydroelectric dams, in *Power Generation and Environmental Change,* Berkowitz, D. A. and Squires, A. M., Eds., M. I. T. Press, Cambridge, Mass., 1969, 79.

54. **Blair, W. F.,** Ecological Aspects, in Water, Man and Nature, A Symposium Concerning the Ecological Impact of Water Resources Development, U.S. Government Printing Office, Washington, D.C., 1972.

55. **Borland, W. M.,** Reservoir sedimentation, in *River Mechanics,* Vol. 2, Shen, H. W., Ed., University of Colorado Press, Ft. Collins, Colorado, 1971.

56. **Brokensha, D. and Scudder, T.,** Resettlement, in *Dams in Africa,* Rubin, N. and Warren, W. W., Eds., Frank Cass, London, 1968.

57. **Burdick, J. C., III and Parker, F. L.,** *Estimation of Water Quality in a New Reservoir,* Department of Environmental and Water Resources Engineering, Vanderbilt University Press, Nashville, Tenn., 1971.

58. **Anon.,** Eutrophication — A Review, California Water Quality Control Board, Sacramento, Calif., 1967.

59. **Chambers, R.,** *The Volta Resettlement Experience,* Pall Mall Press, Garden City, N.Y., 1970.

60. **Farvar, M. T. and Milton, J. P., Eds.,** Shoreline phenomena and their impact on the Nile delta, in *Careless Technology, Ecology and International Development* Natural History Press, Garden City, N.Y., 1972.

61. **Gill, D. A.,** Damming the MacKenzie: A Theoretical Assessment of the Long-Term Influence of River Impoundment on the Ecology of the MacKenzie River Delta, Proc. Peace-Athabasca Delta Symposium, Edmonton, Canada, 1971

62. **Gottschalk, L. C.,** Reservoir sedimentation, in *Handbook of Applied Hydrology,* Chow, V. T., Ed., McGraw-Hill, New York, 1964.

63. **Hutchinson, G. E.,** *A Treatise on Limnology,* John Wiley & Sons, New York, 1957.

64. **White, G. B.,** The New Water Body, in *Symposium on Man-Made Lakes, Knoxville, Tennessee,* American Geophysical Union, Washington, D.C., 1973, 83.

65. **Kuiper, E.,** *Water Resources Development,* Butterworths, London, 1965.

66. **Lowe-McConnell, H. R.,** *Man-made lakes,* Proc. Symp. Royal Geological Society, London, September 30 to October 1, 1965, Academic Press for Institute of Biology, New York, 1966.

67. **Moxon, J.,** *Man's Greatest Lake: Volta,* Andre-Deutsch, Ltd., Accra, Ghana, 1969.

68. **Anon.,** Eutrophication, Causes, Consequences, Correctives, International Symposium on Eutrophication, National Academy of Sciences, Washington, D.C., 1969.

69. **Anon.,** Earthquakes Related to Reservoir Filling, National Academy of Sciences — National Academy of Engineering, Washington, D.C., 1972.

70. **Ruttner, F.,** *Fundamentals of Limnology,* University of Toronto Press, Toronto, 1963.

71. **Worthington, E. B.,** The Nile catchment technological change and aquatic ecology, in *Careless Technology, Ecology and International Development,* Farvar, M. T. and Milton, J. P., Eds., Natural History Press, Garden City, N.Y. 1972.

Part II
River Basin Survey and Assessment

Chapter 6
THE LOWER MEKONG RIVER BASIN*

INTRODUCTION

Floating aquatic plants, such as water hyacinth (*Eichhornia crassipes* [Mart.] Solms), water lettuce (*Pistia stratiotes* L.), and water fern (*Salvinia cucullata* Roxb. *ex* Bory) were observed in large numbers in 1969 in many different locations in the Lower Mekong River Basin. Plants of this type tend to form mats and bogs of intertwined vegetation which are substantially more dense than the growth of the individual species. On this support, many species, including water primrose (*Jussiaea repens* L.) and water morninglory (*Ipomea aquatica* Forsk.), grow and develop. Mechanical and chemical methods of control for these plants have been developed and used to a limited extent in Thailand. Treatment with herbicides of streams, ponds, and tributaries in the area upstream of damsites prior to filling of the reservoir would provide a measure of preventive control. A schedule of annual maintenance application would also reduce the potential growth of aquatic plants above the dam. Natural and artificial irrigation channels should be constructed with level banks and suitable access routes so that mechanical methods of control such as chaining, dragging, mowing, and spraying treatments for aquatic plant control can be efficiently and easily done.

WATER RESOURCE DEVELOPMENT

There are four major river basins in Thailand. The central part of the country is a vast plain devoted mostly to rice cultivation and is irrigated from the Chao Phraya River. The Ping River flows from the northern highlands, and is the major potential source of hydroelectric power at the present time. Irrigation and power development are in progress on the tributaries of the Mekong, but it is the Mekong River itself which has the greatest potential for power and resource development[6] (see Figures 6.1 to 6.6**).

Studies involving technical problems relating to flood control, hydroelectric power production, and water resources development of the Lower Mekong River Basin were begun in 1951 by the United Nations Economic Commission for Asia and the Far East. It was in the context of this effort that the Pa Mong Project was envisioned, and, as a part of the planning program, an aquatic plant control survey to determine the nature and importance of aquatic plant infestations in the area of Pa Mong and ancillary hydrological projects downstream was initiated. It was the purpose of the survey to assess the types of vegetation which might cause maintenance and operational problems, evaluate methods of dealing with these problems, and recommend design modifications and/or further research that might be relevant to the multipurpose water resource projects of the Lower Mekong River Basin. The survey was also intended to take into account physical, biological, ecological, and sociological factors of the environment for their effects on the growth and spread of aquatic plants in terms of economic problems affecting normal operations.

GEOGRAPHIC LOCATION, CLIMATE, SOILS, AND LAND USE

The proposed area of the Pa Mong Reservoir and project lies along both sides of the Mekong River where it forms the boundary line between Laos and Thailand. The damsite is located 20 km upstream from Vientiane, Laos, with one abutment in Laos and the other in Thailand. The Mekong River starts high in the mountains of the Tibetan Plateau and flows in a southerly direction through northwestern Laos, forming the boundary with Thailand for about 80 km on the north and for 720 km on the northeast, and then forms the extensive delta through South Vietnam and finally empties into the South China Sea. The arable lands and potential service areas of the Pa Mong project

This chapter is an abridged and updated version of the unpublished report Potential Growth of Aquatic Plants of the Lower Mekong River Basin, Laos-Thailand, U.S. Army Corps of Engineers, Portland, Ore., 1970.
Figures 6.1 to 6.12 appear following page 52.

include the Vientiane Plain in Laos and vast areas south and west of the damsite extending into Thailand. The area of development has been tentatively estimated to include 1,600,000 ha, of which about 33% of the land is now in rice paddy and upland field cultivation.

The climate of the project area is tropical monsoon. The winters are warm and dry while the summers are hot and humid, the rainy season beginning in mid-May and extending to mid-October. Warm air masses move from the south-west in a counterclockwise direction to produce the rainy season or southwest monsoon. Occasionally, the air masses will shift sufficiently to the west to be displaced from the Indian Ocean, and a period of drought follows. Flooding occurs when the monsoon rains are more frequent than usual at a particular place and time during the rainy season. Cold air masses move from the south over the mainland of China in a clockwise direction during the winter season, mid-October through February, and there is practically no rain. There is a follow-up period of transition when the Pacific air masses of the tropics produce a hot season with sporadic rain.

The soils of the Vientiane Plain and adjacent areas of northeast Thailand are light brown alluvial deposits of fine sandy silt and clay, fairly well suited to rice culture. Under irrigation, two crops per year may be obtained, planted in August and January. Rice yields under present cultural methods are rather poor, averaging about 1.5 to 2 metric tons of rough rice per hectare. Current research indicates that rice yield may be improved greatly by use of new varieties and better management, including greater use of fertilizer, better water control, and control of diseases, insects, and weeds. The upland soils are light sandy loams which tend to be droughty because of the porous subsoil structure, and are very low in plant nutrients and organic matter because of the tropical temperatures and high rainfall. Most of the common vegetables are grown locally. Papaya, bananas, and citrus are popular fruits. Field crops such as corn, sorghum, soybeans, sugarcane, sweet potatoes, peanuts, and tobacco are grown along the river. Pasture crops such as guinea grass, napier grass, coastal Bermuda, and sorghum are grown to a limited extent for grazing for dairy and beef animals. However, fish, seafood, chickens, and pigs are the major source of meat for the Laotian and Thai people.

ECOLOGICAL EFFECTS AND CONSIDERATIONS

Aquatic plant growth involves a very complex biological community, from the most minute plant and animal microorganisms to highly specialized plant and animal life involving a long food chain. The ecological consequences of changes or control of particular populations must be evaluated in terms of the total situation. Both beneficial and harmful effects can usually be cited from the presence or absence of a particular aquatic plant.[14,15,23,25,33]

Water-shading plants are highly objectionable in many situations because they form mats of floating vegetation which interfere with fishing, navigation, and recreational use of water, as well as with the operation of power plants and crop irrigation. Sometimes, however, their removal permits sunlight to penetrate lower strata to encourage bottom growth of submersed aquatic plants that are more objectionable and harder to control than the original surface-covering plants.

Although submersed plants are popularly regarded as beneficial to fish by providing food and oxygen, most modern fishery biologists have observed that extensive aquatic weed growth is detrimental to fish production. Weeds provide sanctuary for small fish resulting in reduced production of larger fish, and they prevent efficient harvesting by impeding netting and boat operations. Rapid death and decay of submerged weeds through natural causes or weed control operations can cause fish kills by oxygen depletion accompanying the decay processes and the evolution of toxic quantities of hydrogen sulfide.

Planktonic algae (phytoplankton) form the basic photosynthetic link between the inorganic constituents in water and higher members of the food chain. When present in sufficient numbers, the color of the water may be green, yellow, red, or black depending upon the species. During the daytime these organisms remove carbon dioxide from the water and produce oxygen as a by-product of photosynthesis. At night, or if photosynthesis is blocked by shading, the process is reversed and oxygen is consumed. Phytoplankton are utilized by certain fish and other marine species and generally increase the yield of fish. Abundant growth of plankton algae with shallow water will shade the bottom and prevent the growth of many submerged vascular plants. How

FIGURE 6.1. Aerial view of the Mekong River as it turns north into Laos toward Luang Phrabang. (Photograph by the author.)

FIGURE 6.2. Narrows in the Mekong River, Pa Mong Damsite, about 20 km west of Vientaine, Laos. (Photograph by the author.)

FIGURE 6.3. Airport at Vientaine, August 21, 1969, flooding on the taxiway from the Mekong. (Photograph by the author.)

FIGURE 6.4. Mekong River, Vientaine, Laos, flooded 9 m above normal. (Photograph by the author.)

FIGURE 6.5. Water hyacinth near Pa Mong on the Thailand bank of the Mekong River. (Photograph by the author.)

FIGURE 6.6. Water hyacinth in the canals in Vientaine, Laos. (Photograph by the author.)

FIGURE 6.7. Water fern, *Salvinia cuculata* Rox *ex* Bory on fisheries lake at Bung Boraphet, Thailand. (Photograph by the author.)

FIGURE 6.8. Close-up of water fern. Some disease and insect damage is apparent. (Photograph by the author.)

FIGURE 6.9. Bladderwort, *Utricularia flexuousa* L., at Bung Boraphet. (Photograph by the author.)

FIGURE 6.10. *Hydrilla verticillata* (L.F.) Casp. with a scum of filimentous algae. Some disease damage apparent. (Photograph by the author.)

FIGURE 6.11. *Monochoria hastifolia* L., a noncompetitive Asiatic species. (Photograph by the author.)

FIGURE 6.12. *Nymphea odorata* Ait., fragrant water lily, a noncompetitive Asiatic species. (Photograph by the author.)

ever, these forms frequently clog water filters, and many kinds interfere with the growing of rice. Most blue-green algae and many green algae produce odors or scums that are undesirable in potable waters and may be allergenic to swimmers. Some species are known to cause gastric disorders to persons consuming the water, and toxic substances are frequently produced that kill fish, birds, and domestic animals.

FLOATING AQUATIC PLANTS

Observations were made of aquatic plants along the Mekong River from Vientiane to Nakhon Phanom, along the Nam Ngum River waterways to the damsite, and over some tributaries of the Mekong River on the Thailand side. The Nam Pung damsite, the Lam Pao Reservoir, and the Ubolratana Dam and Nam Pong Reservoir near Khon Kaen were also surveyed. Field trips were made to the Phayao Reservoir in the vicinity of Chiang Mai, the Bhumibol Dam and Reservoir on the Ping River, and Bung Boropet Lake near Nakhom Sawan. This program was accomplished by use of a Dornier D-80 aircraft for the general survey and a Bell 47G-3Bl helicopter for close inspection and sampling. Motor vehicles and boats were also used as needed. Aquatic plants found during this survey are presented in Table 6.1.

Water hyacinth, water lettuce, and water fern were easily identified from the aircraft by their color, growth habits, and habitats, and large amounts of each were frequently observed during the survey. Dense mats of these covered open-water areas of sloughs, ponds, and other lowland impoundments adjacent to the Mekong River and its tributaries. These areas of weed propagation function as continual sources of drifting vegetation that enter the river during times of flooding. This explained the presence of numerous small colonies of water hyacinth and other floating weeds in the main course of the river at flood stage. Water hyacinth could have come from as far upstream as Luang Prabang, because it is known to grow there as an ornamental plant occasionally fed to swine, as it is in Vientiane. The presence of these weeds upstream of the Pa Mong damsite (e.g., the watersheds of the Mong, Wong, and Loei Rivers) constitute a serious threat to the Pa Mong Reservoir because of their capacity for rapid spread from these isolated infestations during reservoir filling. Easterly winds would keep floating weeds

TABLE 6.1

Summary of Aquatic Plants Observed in Northeast Thailand and Laos According to Type, Common Name and Scientific Name

Common name	Scientific name
Floating plants	
Water hyacinth	*Eichhornia crassipes* (Mart.) Solms
Floating water fern	*Salvinia cucullata* Rox. *ex* Bory
Water lettuce	*Pistia stratiotes* L.
Water velvet	*Azolla pinnata* R. BR.
Duckweed	*Lemna* spp.
Water snowflake	*Nymphoides indica* Kuntze
Water lotus	*Nelumbo* spp.
Submersed plants	
Blyxa	*Blyxa echinosperma* (Clark) Hook
Coontail	*Ceratophyllum demersum* L.
Water fern	*Ceratopteris thalictroides* Brongn.
Chara	*Chara* spp.
Hydrilla	*Hydrilla verticillata* Royle
Limnophila	*Limnophila heterophylla* Berth.
Grassy naiad	*Najas graminea* L.
Ottelia	*Ottelia alismoides* (L.) Pers.
Bladderwort	*Utricularia flexuosa* L.
Marginal plants	
Alligator weed	*Alternanthera philoxeroides* (Mart.) Griseb.
Arundo	*Arundo donax* L.
Coix	*Coix aquatic* L.
Sedge	*Cyperus difformia* L.
Sedge	*Cyperus procerus* Rottb.
Sedge	*Cyperus rotundus* L.
Spikerush	*Eleocharis dulcis* (Brm. f.) Hanschel
Jungle grass	*Echinochloa colonum* (L.) Link
Fimbristylis	*Fimbristylis miliacea* (L.) Vahl.
Lelang	*Imperata cylindrica* (L.) Beauv.
Rounded isachne	*Isachne globosa* L.
Reedgrass	*Ischaemum rugosum* Salisk.
Water primrose	*Jussiaea repens* L.
Pepperwort	*Marsilia crenata* Presl.
Monochoria	*Monochoria vaginalis* (Brn. f.) Prest.
Torpedo grass	*Panicum repens* L.
Water paspalum	*Paspalum scrobiculatum* L.
Giant reed	*Phragmites communis* Trin.
Water smartweed	*Polygonum tomentosum* Willd.
Wild sugarcane	*Sachharum spontaneium* L.
Bulrush	*Scirpus grossus* L.
Cattail	*Typha angustifolia* L.

From Gangstad, E. O., *Hyacinth Control J.*, 10, 4, 1972. With permission.

away from the Pa Mong Dam itself, but any westerly winds, together with river currents, are likely to force mats of floating vegetation against the dam or its protecting boom. Such mats of weeds can become so densely packed that submerged portions decay, slough off, and sink to lower depths where they might enter penstocks and clog turbine screens. Fortunately, floating water fern is not as large and aggressive as *Salvinia auriculata*, the species of water fern that became so troublesome in Lake Kariba, Southern Rhodesia; nevertheless, floating water fern was more frequently observed during the survey than any other floating weed, and it could become most important because it is so widespread. These smaller floating plants along with water velvet and duckweed, which also were observed throughout the survey area, can pass through intake screens and clog the pumps and control structures of irrigation and drainage systems. They are also very troublesome to rice culture.

SUBMERSED AQUATIC PLANTS

Although submersed aquatic weeds could not be identified as to species from the air, dense infestations were observed as dark underwater patches in shallow portions of the reservoirs in northern Thailand (Lam Pao, Nam Pong, and Nam Pung) and in static water sloughs and irrigation channels along the Mekong River and the rice growing areas of Thailand. Submersed aquatic plants are thin stemmed and profusely branched, and most are capable of rapid vegetative propagation from stem nodes, lateral buds, and stolons, as well as by fruits and seeds. They are dispersed by water currents and waterfowl which carry parts of plants and seeds to new areas. They can become established and spread very rapidly as was seen in the relatively new project reservoirs. Once established, they are very difficult to control. Species known to be present in the project area include *Blyxa lancifolia*, hydrilla, coontail, limnophila, grassy naiad, and bladderwort. Bladderwort is regarded as an important food for fish which eat the insect-trapping bladders of the plant. Most of these species grow attached to the bottom mud, but masses of them can drift about underwater with currents and wave action, and they are likely to become prominent in the shallow areas of the Pa Mong Reservoir and in the distribution channels providing irrigation and drainage in rice fields and other agricultural lands. Submersed weeds have been found to impede channel flow and cause flooding as well as crop losses through poor irrigation and water management.

EMERSED AND MARGINAL AQUATIC PLANTS

Tall perennial grasses, such as coix, giant reed, and other herbaceous and woody emersed weeds were observed along the shorelines and banks of all rivers, lakes, and channels surveyed. Bulrushes (*Scripus* spp.), bamboo (*Bambusa* spp.), and other shrubs were also observed among the predominant tall grasses on islands in the Mekong River and on its banks. These rooted, erect plants grow in dense colonies by means of rhizomes, stolons, or runners, and they are capable of rapid encroachment into open water from the margins of lakes or reservoirs covering large areas as they build out from the shore and collect debris among their intertwined root systems. Decumbent creeping species such as water primrose, water morning-glory, and water smartweed were also found to have formed floating islands with mats of floating weeds in the sloughs and impoundments of the area surveyed. Marginal vegetation would not be expected to become serious until after reservoir filling and construction of irrigation and drainage channels of the agricultural distribution system. Seasonal fluctuations in water depths would tend to keep the reservoir perimeter clear of marginal vegetation at first, but aquatic grasses and shrubs are likely to become established eventually along gently sloped margins of bays protected from wave action, as observed at the Nam Pung, Lam Pao, and Bung Borapet Reservoirs in Thailand. Water primrose and creeping perennial grasses such as water paspalum were the most frequently observed emersed weeds growing along the banks, and often across irrigation and drainage channels, in the Ping and Chao Phraya River plains in Thailand, and these or others with similar growth habits are likely to be continuous maintenance problems along the distribution systems of the Pa Mong Reservoir.

AQUATIC PLANT CONTROL

The present survey indicated that aquatic plants generally common to tropical areas around the world[10,11,14,15] are present in substantial numbers along the upper reaches of the Mekong River. There seems to be an unlimited source of these plants in the rice paddies, small ponds, and irrigation channels of the tributaries in the reservoir area. During flood stage it was observed that these plants are uprooted and moved downstream. It is estimated that 10% of the reservoir area is highly infested with these plants, principally water hyacinth. As a measure of preventive control, this area should be treated with (2,4-dichlorophenoxy) acetic acid (2,4-D) immediately prior to filling the reservoir. Spraying by aircraft to reduce the potential population would be particularly feasible at this time because there would be no involvement with crops or probable uses of the water. Water lettuce and water fern, which are also present in local areas, are not readily controlled by 2,4-D at rates commonly used. Spot treatments of water lettuce and water fern with 6,7-dihydrodipyrido-(1,2-a:2',1'-c) pyrazindiinium ion (diquat) at this time would be very effective. Most submersed aquatic plants are also controlled by diquat. Marginal aquatic plants are not controlled by either of these herbicides, but would be controlled in the initial stage by flooding, and would not likely be a problem until after the reservoir had been in operation for some time, and then only in the shallow areas.

PREVENTIVE CONTROL

The principle of preventive control should also be applied after the reservoir is in operation as a repeated maintenance schedule. At the Bhumibol Dam, boats are used to spray water hyacinth that drifts in during the high water and flood stages of the Ping River. According to the Yanhee Electric Authority, which has maintenance responsibility, annual spraying with 2,4-D costs approximately $17,000 to cover approximately 400 ha at a concentration of approximately 2.25 kg of 2,4-D acid per hectare. It is estimated that an annual maintenance schedule for the Pa Mong Reservoir would require treatment of approximately 1% of the surface area.

Other measures of preventive control should be applied to the Pa Mong project if feasible. Clearing of trees and brush around the perimeter of the reservoir and deepening shallow areas will reduce marginal aquatic plant problems. Lining of main irrigation channels will substantially reduce aquatic plant growth in the irrigation system. Tops of channel banks should also be leveled and surfaced so that mechanical methods such as chaining, dragging, mowing, or spraying can be easily and efficiently done. Suitable gates to permit "drawdown" of water in irrigation channels will also facilitate control of marginal and submersed aquatic plants. It may also be economical to fence reservoir areas to permit grazing on marginal aquatic plants by cattle or sheep. Where water buffalo damage the dikes of the channel, these animals should be fenced out.

Various mechanical devices are available for control of aquatic plants, including weed saws, cables, draglines, and barges or boats equipped with mowers, airjets, waterjets, etc. The advantage of mechanical control is that it involves very little direct hazard to fish, wildlife, or humans. However, such methods are usually very inefficient and uneconomical. Usually they give only partial control and must be repeated at frequent intervals. Sometimes mechanical methods only break up or tear the weeds apart and the loosened vegetation, if not properly disposed of, may clog gates, pumps, and syphons, or cause new colonies of weeds to grow downstream. Cables or booms can be used to control water hyacinth and other floating aquatic plants by placing the boom at an angle to the stream flow, floating at the waterline to divert the vegetation to the side where it may be removed mechanically. At the Bhumibol Dam, water hyacinth is pushed to shore by boats or held there by cables at high water level so that it is killed by drying when the water level drops; the stranded vegetation and trash is then burned as soon as possible. Submersed aquatic plants can be controlled mechanically in irrigation and drainage channels by drawing down the water for 2 or 3 days. The thin-leaved submerged aquatic plants dry quickly when exposed to the air. However, some nodes and subsoil propagules commonly survive this treatment and it must be repeated periodically for vegetation control. Some emersed and marginal aquatic plants can be controlled by cutting the vegetative growth at drawdown so that they become submerged and killed when the water level rises again.

CHEMICAL CONTROL

Control of aquatic plants with herbicides is usually easier, faster, and longer lasting, and frequently cheaper, than mechanical control.[9,14, 15,23,25] Treated plants die in place, where they decay and dissipate without causing difficulties in the distribution system. Most of the new aquatic herbicides will not injure crops or fish when used correctly, but some of the older materials such as acrolein or aromatic solvents are toxic to fish. Copper sulfate pentahydrate ($CuSo_4 \cdot 5\ H_2O$) is the safest, most effective, and inexpensive, and most extensively used, algicide. Periodic treatments with copper sulfate at 10- to 14-day intervals at concentrations up to 1.0 ppmw control plankton algae in nonflowing water. Concentrations up to 1.0 ppmw are harmless to most species of fish; higher concentrations are toxic. A problem of algae at the damsite may involve small pipes and fittings in the cooling systems of transformers or other mechanical and electrical equipment. At the Bhumibol Dam, a solution of hydrochloric acid and rust inhibitor is used to clean these parts periodically. Several herbicides are used for control of rooted submerged plants and nonrooted vascular plants, including diquat and 7-oxabicyclo-(2.2.1)heptane-2,3-dicarboxylic acid (endothall). These herbicides are rather expensive and must be applied judiciously. Repeated applications of 2,4-D at 2 to 4 kg of active ingredient per hectare with water carrier provides a satisfactory control of water hyacinth and water lily (*Nymphaea* spp.). Diquat is an effective herbicide for control of free-floating species such as duckweed, water lettuce, and water fern. Rates of 1 to 1.7 kg/ha are sufficient. Applications should be made as a foliar spray with surface equipment in 1650 to 2200 l of water per hectare or with aircraft at 80 l/ha. Most broadleaved, emersed, herbaceous, woody species can be controlled by single or repeated applications of 2,4-D, 2(2,4,5-trichlorophenoxy) propionic acid (silvex), or (2,4,5-trichlorophenoxy) acetic acid (2,4,5-T) at 2 to 10 kg/ha. These include alligator weed, smartweeds, and water primrose (see Table 6.2).

INSECT CONTROL

Techniques for control of aquatic plants with biological agents in tropical climates are relatively well developed at the laboratory level, but few are or have been applied in the operational sense.[3-5,12,14,15] Their application for aquatic plant control in Thailand would require a research and development program on the lakes and reservoirs as they presently exist or may be constructed in the future.

A bagoine weevil, *Neochetina bruchi* Hustache, is being studied in Argentina and the United States. This species is recorded from Guyana, Brazil, Uruguay, and Argentina, feeding only upon species of Pontederiaceae. The adult weevils damage water hyacinth by surface feeding on the foliage. The weevil larvae tunnel and feed in the stem and crown of the plant, young larvae in the petioles, older larvae forming "feeding pockets" in the crowns. There may be from 1 to 12 larvae per plant. A rot usually follows the larval tunneling and the stem and leaf are completely killed. The larvae pupate underwater, forming cocoons from dead root hairs of the plant. *Neochetina* is apparently limited to a completely aquatic environment, although plants nearest the shore receive the most damage.

In laboratory starvation tests, *N. bruchi* adults will feed upon several species of Commelinaceae, and on cabbage and lettuce. This feeding is not extensive and occurs only in the absence of pontederaceous plants; no feeding occurs on these plants if water hyacinth is present.

Neochetina larvae will feed only upon water hyacinth and *Pontederia* in laboratory tests, and the larvae apparently cannot complete their development on other than water hyacinth.

The *Neochetina* weevils would appear to be first on the list to be introduced, and also at this time appear to be the more promising of the insects being considered for importation. Although two species of stem-boring weevils have been found to occasionally feed on water hyacinth in the United States, neither is specific to that plant nor would they offer any competition at all to *Neochetina*. This is not to say that the native *Arzama densa* offers any serious competition to *Acigona infusella;* on the contrary, the South American *Acigona* appears to be needed to complement the North American *Arzama*. However *Acigona* will likely be subject to attack by some of the same parasites presently attacking *Arzama* which of course would reduce the effectiveness of the South American species. These and other species should be studied in Thailand.

HERBIVOROUS FISH

Biological control of aquatic vegetation is possible by use of certain species of herbivorous fish.[1-3,22,36] Fish which feed partly or entirely on aquatic vegetation include Congo tilapia (*Tilapia melanopleura* Dumeril), Java tilapia (*Tilapia mossambica* Peters), Nile tilapia (*Tilapia nilotica* L.), *Tilapia zillii* Gervais, silver dollar (*Metynnis roosevelt* Eig.), *Mylossoma argenteum*

TABLE 6.2

Herbicide, Formulation, General Usage, Application Rate and Chemical Cost per Application

Herbicide (common name) and formulation	General usage or target weeds	Application rate (kg active ingredient per ha)	Application rate (ppm)	Chemical cost per application[a] ($/ha)	Chemical cost per application[a] ($/1000 m³)
2,4-D amine salt (4 lb/gal WSL)	Sedges, broadleaf weeds, and water hyacinth	2–4		2–4	
2,4-D low-volatile ester (4 lb/gal EC)	Same as above including perennials and woody species	2–3		3–4	
2,4-D low-volatile ester (20% G)	Submersed weeds (flooded)	20–40		57–114	
Diquat (2 lb/gal WSL)	Submersed weeds, Water fern, water lettuce	1–1.5	0.5–1		12–24
Endothall, potassium salt (3 lb/gal WSL)	Submersed weeds		1–3		10–30
Endothall, amine salt (2 lb/gal WSL)	Submersed weeds and algae		1–2		13–26
Endothall, amine salt (5% G)	Submersed weeds and algae		1–2		16–32
Silvex, potassium salt (6 lb/gal WSL)	Submersed weeds		1–2		3–7

[a] Prices at 1969 level, shipped F.O.B. from U.S.

Note: WSL = water-soluble liquid, EC = emulsifiable concentrate, and G = granular formulation.

From Nelson, M. L., Seaman, D. E., and Gangstad, E. O., U.S. Army Corps of Engineers, Washington, D.C., 1970.

E. Ahl, common carp (*Cyprinus carpio* L.), the Israeli strain of the common carp, and the white amur (*Ctenopharyngodon idella* Val.). Of these fish the white amur appears to be the most promising for the control of aquatic macrophytes, particularly submersed ones. Local varieties like *Pontius* spp. and some filamentous algae feeders should be studied for control of submerged aquatic weeds in Thailand.

Research on the culture of fish in the irrigation system is currently being conducted at the Agricultural Center, Khon Kaen, Thailand, by Dr. Gerald D. Ginnelly and Mr. Suin Rothcharug. A related project to study the stomach contents of fish is being conducted at Bung Borapet Fisheries Station, Nakhon Sawan, Thailand, by Mr. Camron Phothephitalssa and Mr. Suchin Tongonsee. Basic studies on fish culture are also being conducted by the Department of Fisheries, Gasetsart University, Bangkok, Thailand. These studies suggest that herbivorous fish may be added to the fish population for aquatic plant control and that extensive growth of aquatic plants do reduce the commercial production of fish. From the general review of the different possibilities for control of aquatic plants it is apparent that there is a real potential for biological control, but that practical applications require additional research and development in the target area.

FEEDSTUFF FOR LIVESTOCK

Aquatic plants, including water hyacinth, water lily, water lotus, water chestnut (*Trapa natans* L.) water morninglory, and many others, are used for human and animal food throughout the world. Indeed, food and ornamental uses of these plants account mainly for their wide distribution. Water hyacinth, for example, was brought to Java from its native South America by the Dutch, because its blossoms reminded them of cultivated flowers in Holland. It was introduced from Java into Thailand during the reign of King Rama IV in the 1860s and can be found in the numerous waterways throughout the country. Because it is usually at hand, water hyacinth is used to feed swine, water buffalo, and cattle in Thailand and other Asian countries, and these uses have often been suggested as means of water hyacinth control. Water hyacinth has limited food value (protein content is 8 to 12% of dry weight), but its water content is too high for direct consumption, and the cost of drying and transporting to dryers or feedlots precludes its commercial use for animal feed in quantities sufficient to control its growth and spread. The same would be true of the other noxious aquatic plants such as water lettuce and water fern.

FISHERIES BIOLOGY SURVEY

This study was undertaken as a part of a research program to examine 156 irrigation reservoirs of Northeast Thailand in an attempt to devise methods to improve the culture of fish for local consumption without interference or disruption of the hydrologic purposes of the water storage and irrigation system (see Figures 6.7 to 6.12).

The Huey Syo Reservoir, near Khon Kaen, Thailand, is illustrated in Figure 6.13. Sampling of the water was done at the surface, 1 m below the surface, and on the bottom mud. Sampled species were identified and their potential volume calculated. Bottom sediments were carefully studied in the laboratory.

Observations are summarized as follows: The Huey Syo irrigation system and reservoir located in Nam Pong, Khon Kaen, Northeast Thailand has been in operation since 1951. It has a canal surface area of 76 ha and a volume of 1,666,880 m^3 of water with an average depth of 5 m and a minimum depth of 1 m during the dry season.

The survey of fish species is listed in Figure 6.14. The results indicate a wide distribution of species, some of which have not been reported in this area before. The average standing crop of fish was found to be 4 kg/ha with an F/C ratio of 2:1 and a distribution of 47.6% carp, 9.2% catfish, 26.0% murrels, and 17.2% mixed species. Plankton were found to average 97/m^2 of bottom mud. Most benthos species were Annelides and Chironomids. Gastropods were not frequently observed.

The surface of the reservoir was covered over about 80%. Water hyacinth covered about 40% of the area, water lettuce covered about 20%, and *Hydrilla* covered about 20%. It was concluded that the standing crop of fish was reduced by the excessive growth of aquatic weeds and that herbivorous fish should be studied for aquatic weed control in these reservoirs.

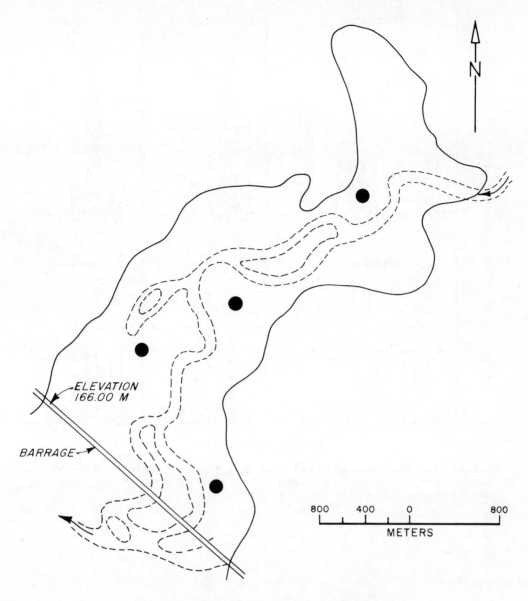

HUEY SYO IRRIGATION RESERVOIR
NAM PONG, KHON KAEN

Capacity 1,166,880 M³

Surface Area 600 Rai = 76 Hectares

Surface Irrigation 5,400 Rai = 864 Hectares

● Survey Point

FIGURE 6.13.   Huey Syo Irrigation Reservoir, Nam Pong, Khon Kaen, Thailand. (From Nelson, M. L., Seaman, D. E., and Gangstad, E. O., U.S. Army Corps of Engineers, Washington, D.C., 1970.)

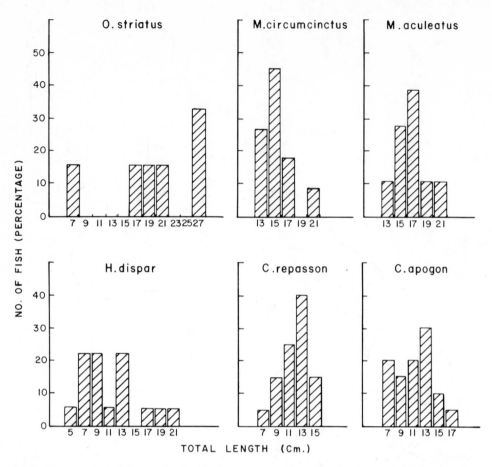

FIGURE 6.14. Length frequency distribution of the economically important fish in the Huey Syo Irrigation Tank. (From Nelson, M. L., Seaman, D. E., and Gangstad, E. O., U.S. Army Corps of Engineers, Washington, D.C., 1970.)

Chapter 7

RIVER BASIN STUDIES IN THE PHILIPPINES*

A survey of aquatic plants in the Republic of the Philippines indicated that water hyacinth (*Eichhornia crassipes* [Mart.] Solms), water lettuce (*Pistia stratiotes* L.), and *Ottelia alismoides* (L.) Pers. were the most prevalent aquatic plants. Frequently, water morninglory (*Ipomea aquatica* Forsk.) and water primrose (*Ludwigia repens* L.) grew intermingled with these species, making the problem of aquatic plant control more difficult. When these aquatic plant growths are permitted to become established in quantity, the original design characteristics of river basins are changed and the storage, drainage, and/or navigation capacity become uncertain. Maintenance operations are required to prevent the problem situation from becoming acute. Mechanical and/or herbicide methods of control are available and should be implemented concurrently with the development of each project; biological control methods should be researched locally for practical long-term effects and methods of control.

RIVER BASIN DEVELOPMENT

Serious flood losses in central Luzon in 1960 dramatized the need for control and regulation of the river systems to insure stability of future agricultural and industrial development. With the endorsement of the Philippine National Economic Council, the USAID/Philippines submitted a project proposal covering seven river systems. Subsequently, a Participating Agency Agreement was executed under which the U.S. Bureau of Reclamation made a survey of water resources[38-44] (see Figure 7.1).

Central Luzon — This area consists of the Agno River Basin and the Pampanga River Basin. The area covers about 18,000 km² in Central Luzon. It is bounded on the south by Manila Bay, on the east by the Sierra Madre Mountains, on the northeast by the Caraballo Mountains, on the north by the Cordillera Central Mountains, on the northwest by Lingayan Gulf, and on the west by the Zambales Mountains. The terrain is relatively flat. It is extensively cultivated and has long been known as the rice bowl of the Philippines. Aquatic weed problems in this area are related to rice culture.

Upper Pampanga River — This project area is located in the province of Luzon on the upper reaches of the Pampanga River Basin. It is planned as a multipurpose project which would furnish year-round water for irrigation of 80,000 ha of land, 30,000 kW of power, and control of flood flows originating above the Pantabangan Damsite, as well as provide municipal water for the area and furnish facilities for fish conservation and recreation. Aquatic weed problems in this project are related to water storage, irrigation, and fish culture.

Cagayan Valley — This basin is located in the northern portion of the Island of Luzon and encompasses part of Isabella, Cagayan Mountain, Nueva Vizcaya, and Quezon Provinces. It covers an area of approximately 28,000 km². The Cagayan River is the principal drainageway, flowing in a northerly direction from its headwaters in Nueva Province. The principal land forms are the Sierra Madre Mountains on the east, the Cordillera Central Mountain Range on the west, and the Caraballo Mountains on the south. The river valley is a broad fertile alluvial plain, subject to periodic flooding during high water flows. Marshes and swampland are found along the lower reaches at the mouth of the river. Numerous potential dam and reservoir sites are available along the watercourse. Aquatic weed problems are related to flood control.

Bicol Penninsula — The basin is situated on the lower part of the Island of Luzon and encompasses the Bicol River in the Provinces of Carmarine Norte and Arby. It is an elongated flat plain bordered by mountains and volcanos on the eastern side and highlands and low hills on the western side. It covers an area of about 312,000 ha. The main drainage is to the northeast through the Bicol River and its tributaries which ultimately drain into San Miguel Bay. There is only limited opportunity for large-scale hydroelectric power development. The most pressing need is for the expansion of irrigation storage reservoirs. Aquatic weed problems are related to water storage.

Negros Island — The Ilog-Hilabangan River

---

*Abridged and updated from unpublished report, "Potential Growth of Aquatic Plants in the Republic of the Philippines and Projected Methods of Control," Office of the Chief of Engineers, Washington, D.C., 1971.

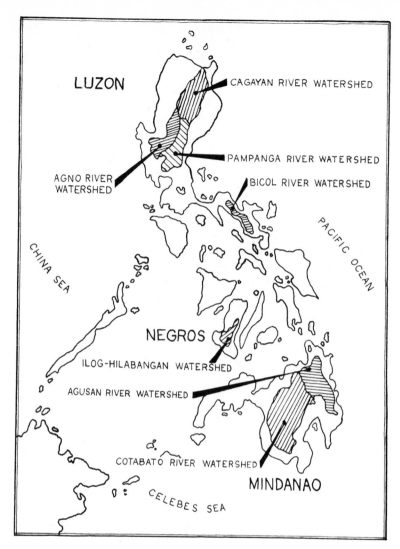

FIGURE 7.1. Location map of watershed studies in the Philippine Islands. (From Gangstad, E. O., *J. Aquatic Plant Management,* 14, 10, 1976. With permission.)

Basin is located in the southern part of the island, bounded on the north by the Panay Gulf, on the south by Tolong Bay, and on the east by Tonon Strait. The Ilog River, the principal drainageway, discharges into the Panay Gulf. The Hilabangan River originates in the easternmost part of the island and flows in a westerly direction into the Ilog River, about 25 km from the mouth. The basin is characterized by a dry season, January through May, and a wet season, May through December. Flooding is a serious problem and would be a major justification for the project. Aquatic weed problems would be related to flood control.

Aguisan River — This river basin is in the eastern part of the island of Mindanao, the second largest island in the Philippines. It includes most of

Aguisan Province and a large part of Davao Province. The basin covers about 11,500 km². The Aguisan River is the principal drainageway. It originates in Davao Province and flows in a northerly direction, and discharges into Butan Bay. Numerous potential damsites exist but basin engineering data are not available to formulate a comprehensive plan. Aquatic weed problems would be related to flood control.

Cotabato — This basin covers an area of approximately 20,000 km² in eastern Mindanao. It is bounded by the Central Cordillera Mountains on the east, the Tiruray Highland in the south, and the Lanao-Bukidon in the north. The principal drainageway is formed by the Pulangi and Alah Rivers. The Liguasan and Libugan Marshes cover extensive portions of the lower basin. Lake Bulan and Lake Sebu are two large bodies of inland water. Much of the arable land is periodically flooded during the rainy season, and flood control would be a major justification for construction of the project. Aquatic weed control would be related to flood control.

AGRICULTURAL DEVELOPMENT

Agricultural production in the Philippine Islands is largely limited to a pastoral system of the small family farm. Unit production is low and unit costs are high, because much of the field work is done by hand, crop varieties are low yielding, market facilities are generally inadequate, and input costs are high for most elements of the production system. The Filipino farmer sells his crop at a low competitive price and he must buy fertilizers, pesticides, and machinery at high import prices. As a result, agricultural production does not meet the needs of the people for food and fiber.[7-9]

Programs to stimulate production of agricultural commodities to meet current needs for food and fiber are in progress. The International Rice Research Institute has a program of development of high yielding varieties. Similar programs are under way at other universities and colleges. The Department of Agriculture has programs for better utilization of fertilizers and pesticides. The program includes extended use of irrigation water, and an effort is being made to improve marketing of agricultural products.

Studies of river basin development suggest that practically all phases of agriculture and local industry could be expanded under a program of development of water resources. Inasmuch as water supply itself is not limiting, there is reason to believe that such a development would be in the interest of the public. Philippine agriculture is neither as labor-intensive as that of other countries in Asia nor as capital-intensive as that of western countries. The overall picture of low productivity per man-hour of labor, as well as the meager yields per hectare, are nevertheless real. While abundant rainfall and high temperatures sufficient for a long growing season are important assets to be taken into account for future development by the addition of improved technology, these same features rob the soil of nutrients and require improved management to attain these ends.[7,9,16]

SURVEY OF AQUATIC PLANTS

Aquatic plants are commonly found in all bodies of water in the Philippines to a greater or lesser extent. The diverse nature of these habitats provides for a great many different ecological adaptations, but from the standpoint of control, these plants are simply classified as floating, submersed, and marginal. Floating plants may be free-floating or rooted and emergent. Submersed plants may be free-floating or rooted. Marginal aquatic plants are equally diverse, ranging from those rooted to the bank and growing horizontally out over the water to those which grow upright on the shoreline.[9,14,15] A summary according to plant type, common name, and scientific name is given in Table 7.1.

Floating aquatic plants — Water hyacinth, water lettuce, and water fern were the most widely observed floating aquatic plants in the Philippines. Water hyacinth is by far the most troublesome pest. Under the influence of current, wind, or tide, masses of water hyacinth move about in rivers, lakes, reservoirs, and navigation channels and increase the hazards of operation. They interfere with the movement of all types of craft. These plants may accumulate in huge masses at the bow of a barge or boat and may even stop forward motion. Floating masses may also interfere with the operation of locks and gates involved in a water control distribution system. They are known to build up in sufficient quantity to destroy a bridge and/or related structure of the waterway.

Submersed aquatic plants — Vallisinaria, hydrilla, and ottelia were observed to be the most troublesome submersed aquatic plants. Frequently the latter is found growing with floating types and presents a difficult problem for control. Submersed plants interfere with small boats in a closed channel by jamming the propeller, clogging the cooling system, blocking the rudder, and possibly stopping all forward motion of the boat. These plants may retard the flow of water as much as 80%, reducing the irrigation system to a point of uselessness. Extensive growth may also interfere with the development of fisheries and limit industrial and municipal uses of water.[7,9,26]

Marginal aquatic plants — By definition, these plants grow along the shoreline and are a natural part of the vegetation. They include an extremely diverse group of plants from dicots to sedges and aquatic grasses to woody plants (such as the water

TABLE 7.1

Common name	Scientific name
Floating and floating leaf	
Water hyacinth	*Eichhornia crassipes* (Mart.) Solms.
Water lettuce	*Pistia stratiotes* L.
Water lily	*Nymphaea* spp.
Watersprite	*Ceratopteris siliquosa* (L.) Copel.
Cloverfern	*Marsilea crenata* Presl
Water fern	*Salvinia* spp.
Submersed	
Ottelia	*Ottelia alismoides* (L.) Pers.
Hydrilla	*Hydrillia verticillata* Royle
Elodea	*Elodea* spp.
Myriophyllum	*Myriophyllum* spp.
Curlyleaf pondweed	*Potamogeton crispus* L.
Marginal	
Water willow	*Ludwigia octovaluis* var. Sessiliflova (Mich) Raven
Water primrose	*Ludwigia perennis* L.
Water morninglory	*Ipomoea aquatic* Forsk.
Kalagoa	*Monochoria vaginalis* (Burm.) Presl
Veronica	*Veronica cinerea* (L.) Less
Maismais	*Sphenochlea zeylanica* Gaertn.
Jungle grass	*Echinochloa colunum* (L.) Link
Barnyard grass	*Echinochloa crusgalli* (L.) Beauv.
Bagang	*Phragmites australis* (Cav.) Trin.
Sedge	*Cyperus difformis* L.
Sedge	*Cyperus compactus* Retz.
Sedge	*Cyperus imbricatus* Retz.
Sedge	*Cyperus iria* L.

From Gangstad, E. O., *J. Aquatic Plant Management,* 4, 10, 1976. With permission.)

willow). In the Philippines, marginal aquatic plants such as water primrose and the water morninglory present the greatest problem. They tend to overgrow a particular site and render it useless for domestic purposes. If the banks are properly engineered, however, the weeds frequently can be controlled by mechanical means, and the foliage can be used agriculturally.

MEASURES OF CONTROL

Control measures to date in the Philippines have been largely limited to mechanical devices such as floating booms, draglines, and manual removal of plants by hand and hand tools. In the United States the same methods have been used, and specialized equipment has been developed by commercial companies and the Corps of Engineers. For water hyacinth control, the most effective mechanical method is the "Destroyer" or "saw-boat." Across the bow of this boat is mounted a horizontal axle with cotton-gin saws spaced about 10 cm apart, raised and lowered on an outrigger. For this method to be effective, it is usually necessary to cut the floating aquatic material four to five times. The chopped material decays and falls to the bottom in 3 to 5 weeks. The boat developed by the Corps of Engineers is illustrated in Figure 7.2. The sawgang is driven about 1000 rpm. When the boat is in operation, the action of the saws will provide forward thrust, but an auxiliary barge is used to move the machine at a steady pace and to move it from place to place.

Chemical Methods

Most aquatic plant control done since the end of World War II has been accomplished by means of herbicides. By far the larger part of this effort has been with the herbicide (2,4-dichloro-phenoxyacetic acid, 2,4-D), and most of the treatment in aquatic sites has been for control of water hyacinth. Herbicide may be applied from the bank or shoreline, from a boat, or from an airplane. Most of the work done by the Corps of Engineers has been with a standard boat using pumps of 38 to 132 l capacity at 11 to 54 kg/cm^2. There are a number of innovations that can be introduced to reduce drift of the herbicide. These should be used as needed (see Figure 7.3).

Biological Control

Techniques for control of aquatic plants with biological agents in tropical climates are relatively well developed at the laboratory level, but few are or have been applied in the operational sense. Application of these agents for aquatic plant control in the Philippines would require a research and development program on the lakes and reservoirs as they exist or may be constructed in the future.

Biological control by the use of the alligator weed flea beetle (*Agasicles hygrophila* Selman and Vogt) to control alligator weed in the United States is one example of the successful use of an organism to control aquatic plant growth. However, alligator weed was not observed in the Philippines. Currently, a number of insect enemies

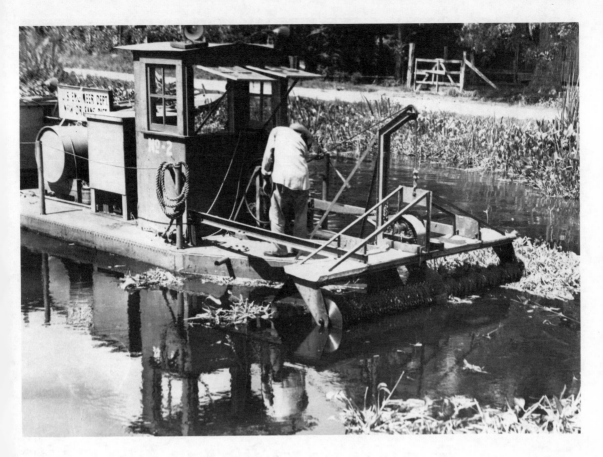

FIGURE 7.2. A workman adjusts the height of the saw on the New Orleans District's Destroyer No. 2. The circular blades mutilate water hyacinth for mechanical control. The boat has not been in use since the use of 2,4-D began in 1947. (Photograph courtesy of New Orleans District, U.S. Army Corps of Engineers.)

of water hyacinth are being studied, but techniques are not yet operationally developed for application in the Philippines.[11,19,20,22]

Use of herbivorous fish to control aquatic vegetation has received considerable attention. Fish which feed partly or entirely on aquatic vegetation include Congo tilapia, Java tilapia, Nile tilapia, silver dollar, common carp, the Israeli strain of the common carp, and the white amur. Of these fish the white amur appears to be the most promising for the control of aquatic macrophytes, particularly submersed ones[1,2,8,13,17,18,22] (see Figure 7.4).

The white amur is native to those rivers of China, Manchuria, and Siberia which run into the Pacific Ocean from latitudes 50 to 23°N. This fish has been successfully introduced into a number of countries in Southeast Asia, Eastern and Western Europe, and the United States for research and field trials. Arkansas is the only state in the United

States to have released the white amur in natural bodies of water.[34]

The white amur is also called the grass carp or Chinese grass carp. It is classified as part of the Cyprinidae family, to which the carp and minnows belong, and can be readily distinguished from other carp by its double-rowed, compressed, and comb-like pharyngeal teeth. The pharyngeal teeth are falciform and toothed. The body of this fish is slightly elongated and is moderately compressed laterally. The upper part of the body is dark gray to olive brown and golden brown with the lower part silver-white in color. The white amur has a high tolerance to temperature changes and can withstand salinities as high as 10,000 ppm and oxygen concentrations as low as 0.5 ppm. Only two instances of natural spawning outside their native range have been reported; however, artificial spawning of the white amur can be conducted with relative ease. Artificial spawning could be

65

FIGURE 7.3. Airboat spray operations. The airboat has the advantage of maneuverability to spray in shallow waters. (Photograph courtesy of New Orleans District, U.S. Army Corps of Engineers.)

used to control the number of fish in an area where natural reproduction does not occur.[1,13,18,35]

Growth of the white amur depends on the amount and type of vegetation present, length of the growing season, temperature, and the inter-relationships of these factors. The white amur prefers soft vegetation and will eat more than its weight daily of such plants as pondweeds (*Potamogeton* spp.), coontail (*Ceratophyllum* spp.), elodea (*Elodea* spp.), and cattails (*Typha* spp.). Hydrilla is readily consumed by this herbivorous fish, with the amount eaten depending on water temperature and fish size.[35-37]

The white amur or grass carp spawns naturally in its habitat, the rivers draining to the Pacific between the Amur River, in about latitude 50°N, and the West River, or Si Kiang, in about latitude 23°N, and in all the rivers between. It has also

spawned naturally in some of the countries to which it has been transplanted, namely, Taiwan, Japan, the Ukraine, and possibly in the Moscow district. Until about 1962, the commerical supplies of the fry of this valuable fish for stocking in fish farms, chiefly in the countries of southeast Asia, came from the Chinese rivers, and especially the Yangtse Kiang and the West River. A considerable industry there was concerned with the catching, sorting, rearing, and distribution of naturally spawned fry caught in the rivers.[13]

It has long been well known that, if the major Chinese carps are kept in ponds until they are of a size to mature, they will develop mature gonads. But though the males will produce ripe sperm, the ovarian eggs of grass carp are, under captive (pond) conditions, not transformed from the fourth into the fifth phase of development. In natural conditions this change takes place, and fertile eggs are

FIGURE 7.4. White amur or grass carp, approximately 6 years old, 650 mm in length and 158 mm in width, weighing 5.78 kg. (Photograph courtesy of Dave Sutton, University of Florida.)

ovulated. But female grass carp can be induced to ovulate fifth-phase eggs under pond conditions if "estrualized" with extracts of the pituitary glands of fish, thus to some degree replacing whatever natural stimuli bring this change about in the spawning grounds.[13]

The artificial inducement of spawning in the grass carp (and the other major Chinese carps) is now well established in several countries, and the way is clear to integrating these valuable and useful fish into the fish culture systems of these and other countries. There is a considerable variation in the techniques used and in the dosages given. This suggests that there is wide margin of error, provided that the female recipient fish is fully ripe (no difficulty is recorded for the male). Uncertainty in this regard remains the chief difficulty and probably the main reason for unsuccessful treatments (see Table 7.2). The spawners may be held immediately before and after treatment in small ponds with flowing water, or in small cement tanks with flowing water. Flowing water stimulates the fish to increase the number of eggs produced, and also increases the percentage of eggs fertilized.

Pituitary glands from the donor fish may be used fresh, or preserved in absolute alcohol or acetone, which also removes the water and fat from the gland. Fresh glands are preferred wherever possible, as most workers find them more effective than preserved glands. In India experiments show that extracts of fish pituitary glands in glycerine and in trichloracetic acid (T.C.A.) retained their potency for many days under refrigeration, and, in the case of glycerine extracts, even at room temperatures.[13]

TABLE 7.2

Dosages of Fish Pituitary Glands Used to Induce Spawning of Grass Carp, and the Results Obtained

Country	Donor fish	Dosage given to female grass carp	Results
China	*Cyprinus carpio*	4 to 10 glands per recipient fish	33% successful injections
China	*C. carpio, Carassius auratus,* and other fish	3 to 3.5 glands per kg or 2 to 3 mg dried defatted glands/kilo	Mass-production of fry; 1200 million in 1962
Taiwan	*C. carpio*	1 to 1.5 glands from donor fish of weight equal to that of recipient	28% successful injections
Taiwan	*C. carpio*	Glands from donor fish of weight 80–100% or more of weight of recipient fish: or 3 to 4 mg/kg of dried defatted glands	Early in season nearly 100% successful; July 40%; August– September 25% successful
Ukraine	*C. carpio*	1 gland per kg of recipient fish	Successful, but few young
Turkmenina	*C. auratus, C. carpio*	1 to 7 glands per kg of recipient fish	2 out of 3 injections successful, 7–9000 fry
Russia	*C. carpio*	In recipient fish of 6 to 12 kg, a dose of 2.5 to 3 mg dried defatted glands, followed by 2.5 to 3 mg/kg	Mass-production of fry; 90 million Chinese carp fry in 1964–65

From Lin, S. Y., Chinese-American Joint Commission on Rural Reconstruction, Fisheries Serial 5, 1965.

Chapter 8

ESTABLISHED IRRIGATION SCHEMES IN INDIA*

INTRODUCTION

Serious aquatic plant infestations in India include water hyacinth (*Eichhornia crassipes* [Mart.] Solms) and hydrilla (*Hydrilla verticillata* Royle). These and other aquatic plant infestations severely hamper agriculture that is dependent on irrigation. High populations of aquatic weeds seriously limit the flow of water, interfere with structures necessary for the transport and distribution of water, and infest cultivated areas in the field. Present control methods are largely mechanical, involving mostly hand labor. Although there is no apparent lack of hand labor, these methods are not meeting the demands of the irrigation schemes. More efficient mechanical methods, improved herbicide treatments, and extended use of biological controls are necessary for water resource development and utilization.

The Republic of India — Under provisions of the Indian Independence Act of 1947 of the British Parliament, the States of India and Pakistan were established, and their freedom became effective August 15, 1947. The Republic is comprised of sixteen states and six centrally controlled territories. The capital is located in New Delhi, the northernmost large city. India is one of the most populated countries of the world, with a total population estimated at 610,000,000 in 1976.

Physical characteristics — India is naturally divided into four geographic regions, the Himalaya, the North River Plains, the Eastern Ghats, and the Western Ghats, The Himalaya region is a complex mountain system about 160 km wide which extends for about 2500 km along the northern and eastern borders. It is part of the highest mountain system of the world and is a region of extreme cold and rugged terrain of little agricultural or industrial value. South, and parallel to this region, lies the North River Plains region, a belt of alluvial flatlands, about 300 km in width. This region is watered by the Indus, Ganges, and Brahmaputra Rivers and includes a large part of West Pakistan, and East Pakistan to the Assam. The western portion of this region is watered by the Ganges River and is known as the Gangetic

Plain. The eastern portion of this region is watered by the Brahmaputra River. This is the most fertile and highly populated area of India. It is a rocky tableland occupying most the area of peninsular India. It is divided by low mountain ranges, 300 to 1000 m elevation. The Western Ghat is an escarpment of land extending to the Arabian Sea at an elevation of about 1000 m. The Eastern Ghat is a rolling flatland extending to the Bay of Bengal at an elevation of about 500 m, including a narrow coastal plain known as the Coromandel Coast. There are three main rivers in the area, the Cauvery, the Kistna, and the Godavari Rivers.

Climate — Because of the diversity of the topography and the geographical position of the subcontinent peninsula, there is a great diversity of climate in the Republic of India, ranging from a north temperate climate along the slopes of the Himalaya to south temperate, subtropical, and tropical climates in the central and coastal plains. Seasonal variations in rainfall result from the southwest and northeast monsoons. The rainy season extends from June to November, due to moisture-laden winds blowing across the Indian Ocean and the Arabian Sea. Precipitation includes a high of 300 cm or more along the slopes of the Western Ghats. In the northeast section, rainfall occasionally exceeds 1250 cm per year. Mean annual rainfall of the Himalaya area is about 150 cm per year. Precipitation on the central plateau, a major portion of the Republic, is only 40 cm per year. During the dry season, December through February, cool weather prevails over most of the subcontinent. The hot season, March through May, may include temperatures upwards of 35 to 45°C.

Vegetation — There is a wide diversity of plants in India associated with extreme differences in climate. In arid areas of West Pakistan, the flora is sparse and mostly herbaceous, including thorny species, bamboo, and some palms. In the more humid areas of the Gangetic Plain, mangroves flourish at the lower elevations and tropical hardwoods at the higher elevations. Coniferous species, including pine and cedar, predominate in the northwest portion of the Himalaya region. Extensive tracts of impenetrable jungle are found

*Abridged and updated from unpublished report, "The Potential Growth of Obnoxious Plants and Practical Systems of Control in the Republic of India," Office of the Chief of Engineers, Washington, D.C., 1971.

in the swampy areas of the Western Ghats. Vegetation is much less abundant in the central plateau, in some places almost nonexistent.

CROP IRRIGATION AND WATER RESOURCES

India is known to have had irrigation schemes early in the history of the country, and with the possible exception of small rivers and streams high in the Himalaya, there are no rivers which could be described as natural or "wild." Down through the ages, various irrigation projects have been successfully used. However, it is only within recent times that hydroelectric power and multiple use concepts have been introduced into the system. At the request of the Agency for International Development to briefly survey irrigation and water resources in India, visits were made to: (1) the Commonwealth Institute of Biological Control at Bangalore, (2) the Mysore Engineering Resource Station at Krishmaraza Sagar, (3) the Tungabhadra Project Dam and High Level Canal Scheme near Hampi, India, and (4) the Ministry of Irrigation and Power at New Delhi, India. The report also includes a summary of the technical observations made during the survey of the Chambal Command Area of Rajasthan, India.

AQUATIC WEED CONTROL PROBLEMS

Several species of aquatic weeds have infested the inland waters of India including lakes, ponds, tanks, irrigation channels, and cultivated areas temporarily covered with water, such as in jute or rice culture. Some of these species, such as water hyacinth,[28,29] alligator weed,[21] water lettuce,[12] and water primrose (*Jussieua repens* L.),[29] are introduced species and have caused enormous losses. They are transported inadvertently from place to place and may prove to be more serious in years to come. Control of these species has been largely neglected. Biological control holds great promise through the introduction of biotic agents or organisms from other parts of the world.[1-3, 5-7,10,11,24,27,28] Some weed insects and plant pathogens are currently under study in the U.S. with the view of introducing them to India and/or other parts of the world for biocontrol of aquatic weeds. Studies on aquatic weed control by the Irrigation Research Institute, Roorkee, India,

indicate that weed-infested sites have wide differences in species, density, and other environmental factors. Chemical control that has been quite successful in other countries has not been sufficiently successful in India for general recommendation.

THE TUNGABHADRA PROJECT DAM AND HIGH LEVEL CANAL

This project is administered by the Minister of Irrigation and Power, New Delhi, India. The Tungabhadra River is about 640 km in length and joins the Krishma River at Sangameswaram in Andhra Pradesh. The major portion of the river is in Mysore State. The total catchment of the river is about 45,000 km². Even though the monsoon flows are large, the summer flows dwindle greatly, so that water storage is an important part of the scheme. The dam is of medium size, i.e., a main dam in masonry across the river 1700 m in length with an earthen dam of 1500 m on the left side and a composite dam of 687 m on the right.

Project Origin

Pampa was the ancient name of the Tungabhadra River and it still survives in the name of the village of Hampi originally mentioned as "Pampa Tirtha" in the mythic Ramayana. The river Tungabhadra derives its name from its two tributaries, viz., Tunga and Bhadra, both of which rise in the same hill, Varaha Parvata in the Western Ghats in the Mysore State. After running widely different courses, they unite at the sacred village of Kudali.

The major portion of the river can be reckoned as running through the Mysore State, as it traverses only 136 km in Andhra Pradesh. Prior to the reorganization of the states, this river used to flow through four states, viz., Mysore, Bombay, Hyderabad, and Madras.

The total catchment of the river is 42,2690 km² and at Mallapuram it is 17,408 km². The gradient of the river bed varies widely and forms a series of rapids. The important rapids occur when the river cuts through the Sandur hill range above Mallapuram and the other range of hills near Hampi 7 mi lower down where the river flows through the historic ruins of the capital of the famous Vijayanagar Empire.

There are no major tributaries to the Tungabhadra. The noteworthy minor ones are the

Varada, the Bedavati or Hagari, and the Hindri; the first joins the river above Mallapuram and the others below Mallapuram. The river traverses through a terrain under the influence of the southeast monsoon and receives a copious supply from its highly wooded and hilly catchment in the Ghats. The maximum and minimum yield of the catchment is estimated at 1.7 million hectare-meters (1 ha-m = 8.108 acre-feet) and 0.8M ha-m respectively, the average being 1.2 million ha-m. The average rainfall in the catchment is about 100 cm/year, maximum being 400 cm in the Ghats and minimum 850 cm in the plains.

Past Irrigation Schemes

Since ancient times the water of the Tungabhadra River has been used for irrigation. During the reign of the Vijayanagar Kings, more than three centuries ago, various anicuts of massive size were built across the river to the east and west of Vijayanagar. From these anicuts, irrigation channels took off on both sides of the river. These irrigation canals have been built and rebuilt many times.

The Tungabhadra River Project

The government of Madras approached the government of India in 1935 for appointment of an Inter-State Commission to give a decision on the equitable sharing of waters by the four governments of Madras, Hyderabad, Bombay, and Mysore. Later on in 1936, the governments of Mysore and Madras met and came to an agreement. In 1938, the Madras and Hyderabad governments conferred and arrived at an agreement for partial utilization of the available supply to the extent of 2.8 thousand million hl each for new irrigation and 0.4 thousand million hl extra to Hyderabad for power generation, which after delivery to the river could be utilized by Madras for new or old irrigation.

The four districts of Bellary, Kurnool, Ananthapur, and Cuddapah, known as ceded districts, and to a lesser degree Nellore, had in the past been cruelly affected by successive famines and scarcity due to deficient or ill-distributed rainfall. The need for protection of the area was imperative in order to conserve the life and fertility of soil. It was to afford some protection to such a badly affected famine region that an irrigation storage project on the Tungabhadra, capable of delivering water to a large extent of land, was considered absolutely necessary.

The scheme included a high and low level canal. The high level canal is kept at 535 m and the low level at 475. On the left side of the river the scheme provides for a canal in Raichur District and development of power at the foot of the dam and at various drops along the canal. The present scheme comprises a dam for impounding 37 thousand million hl of water with the right bank canal, 315 m in length, running in the districts of Bellary and Kurnool, and the left bank canal, 200 km in length, running in the district of Raichur.

Due to the formation of Andra State on October 1, 1953, the Tungabhadra Board was formed by the government of India to be in charge of the following: (1) half the Tungabhadra Dam on the right side, (2) maintenance of the Right Bank Low Level Canal in the common portion, (3) execution of the High Level (Main) Canal in the common portion (up to the Mysore-Andhra Pradesh border) and (4) the Dam Power House on the right side, and the Hampi power house.

Right Bank High Level Canal

The High Level Canal Scheme was inaugurated by Shri Kadidal Manjappa, Chief Minister of Mysore, and Shri C. M. Trivedi, Governor of Andhra Pradesh, on October 2, 1956.

Salient Features of the Scheme

The scheme provides for utilization of 2.8 thousand million hl water from the reservoir from June to November through the vents of the High Level Canal (ten in number), to discharge a maximum of 4000 ha-cm/hr. Out of this, 1385 ha-cm/hr will be utilized for irrigation in the Bellary District and 2585 ha-cm/hr in the Ananthapur and Cuddapah Districts. This is mainly a single crop scheme, partly for the irrigation of dry crops and partly for the irrigation of wet crops.

The High Level Canal will be a contour canal up to the regulator beyond the Pedda Hagari crossing, where the Guntakal Branch and the outfall into the Pennar will branch off. The water will be dropped into the Pennar near Uravakonda and will be picked up by a system of canals proposed to branch off from the Pennar River at the Mid-Pennar Regulator and Gandikota weir. The first 40 km of the canal will run in rugged country involving long and deep cuttings, side walling, and finally a tunnel between the fortieth and forty-first kilometers. After crossing the Bellary-Rayadurg Railway line, the canal will be taken across the Chinna Hagari and Pedda Hagari Rivers by means of regulators. The High Level Canal is proposed to be lined throughout up to the regulator beyond the Pedda Hagari crossing in order to minimize seepage and ensure the safety of the canal.

The canal, after cutting the Daroji hill range by a tunnel, passing the hillocks by deep cuts, and crossing the major valleys by tunnels and syphons near Bhanusandra, Nari Halla, and Karigana Halla, passes the ridge near Bellary town and enters Andhra Pradesh. Thereafter, the canal crosses the rivers Chikka Hagari and Pedda Hagari by aqueducts, and the main canal drops into the Pennar River through the watershed cutting at the Uravakonda deep cut. The water is again picked up at the Mid-Pennar Regulator near Penakacheria and two canals, Mid-Pennar North Canal and Mid-Pennar South Canal (one on the right side and the other on the left side) were excavated to irrigate the Anantapur District as part of Stage I. The waters of the Pennar also mingle with the waters of the Tungabhadra High Level Canal, and the same is again picked up at Gandikota by a weir and two canals, Cuddapah North Canal and Cuddapah South Canal, which are proposed for feeding the Cuddapah District as part of Stage II. A branch canal, Guntakal Branch, will divert water from the High Level Canal in Andhra Pradesh to irrigate 24,970 ha.

Construction

The dam begun in 1945 is complete in all respects and water is being stored. The Left Bank Canal up to Kilometer 203 and the distributaries served by it has been completed and water let out. Further extension of the canal is in progress. Out of the total area of 232,000 ha under the canal, a potential of 207,277 ha for irrigation agriculture has been created and an area of 157,002 ha is utilized.

The Right Bank Low Level Canal and the distribution system in both the States of Mysore and Andhra Pradesh have been completed. As against the area of 36,938 ha for irrigation under this canal in Mysore, the potential created is 36,572 ha and the utilization is 31,380 ha. As against the area of 69,490 ha for irrigation under this canal in Andhra Pradesh, the utilization is 61,700 ha.

The first stage of the main canal, the Right Bank High Level Canal, up to the Mysore border under the control of the Tungabhadra Board is practically completed as planned. The construction of the distribution system by the respective state governments is in progress.

The successful completion of High Level Canal Scheme Stage I will bring relief to the chronic famine-stricken zones of Rayalaseema. As the waters of the Tungabhadra flow, they may be expected to usher in a new era of prosperity.

In the words of the late Prime Minister, Shri Jawaharlal Nehru,

These big modern dams and canals are our Modern Temples. The Tungabhadra Project with its canals, power houses and fisheries, is one such Temple, a gift by the thousands of workers and engineers, those who left their hearth and home to work for the benefit of their fellowmen including those who gave up their lives in this service and did not live to see and enjoy the fruits of their labour.

Development of Fisheries

With the construction of the dam across the Tungabhadra River, a vast reservoir spreading 379 km^2 in area, and 80 km in length has come into existence. This reservoir is ideally suited for growing fish, which will not only increase the nutritional standards of the people but also fetch a considerable amount of revenue. A fish farm is the main center of activity at present. This farm has gradually expanded, now containing 80 cement cisterns and 40 earthen ponds.

Induced breeding of major species of carp by administering pituitary injection to the parent Gangetic carp was attempted, and the unit has succeeded in producing fish seed at a level which has reached 218 million seedlings of major species of carp in 1966 to 1967. The breeding of common carp has also been taken up, and the production of seedlings is gradually on the increase. The success in this direction has obviated the need for the import of fish seed by the Mysore Fisheries Department from the Fish Seed Syndicate, Calcutta, which had previously been air-lifted at considerable expense. This fish farm is now an important source of supply of fish seed to Mysore, Andhra Pradesh, and Pondichery.

THE CHAMBAL COMMAND AREA IRRIGATION SCHEME

Operational Plan

The Chambal irrigation scheme began operating

in 1960. It was designed to irrigate 2.3 million ha. This acreage is equally divided between the states of Rajasthan and Madhya-Pradesh. The Chambal Command Area contains the Chambal River, which enters the Kota District from the western side, and the Parbati, Kalisindth, and Parwan Rivers which drain the excess water out. Rainfall in the area averages 77 cm a year with most of it received during a 40-day monsoon.

Water is stored in a series of dams along the Chambal River. These dams are the Grandi Safar, Rana Pratrap Sagar, Jawahar Sagar, and Kota Barrage. Total water storage area is 563 km^2. Considering the entire irrigable acreage, this will provide 90 cm of water per year when evaporation and distribution losses are included. The rapid increase in irrigated area is beginning to demand maximum flow through the main canal system.

The Kota Barrage was constructed for the purpose of raising the water level to feed a canal on the left and the right of the central Chambal River. The left bank canal, which is only in Rajasthan, has an inflow capacity of 1270 ha-cm/hr and a total distribution system of 960 km (excluding field channels).

The right bank canal has an inflow capacity of 6650 ha-cm/hr, of which 3900 ha-cm/hr is proposed for delivery to Madhya-Pradesh. This canal is 320 km long. The total distribution system is 1280 km long in Rajasthan (excluding field channels). The 2750 ha-cm/hr available for Rajasthan will irrigate 176,000 ha.

Serious problems of waterlogging, soil salinity, and aquatic weeds in the distribution system have arisen since it began operation in 1960. The water table rose rapidly in the surrounding area after water was released into the canals. The government of India, realizing drainage was a very necessary part of irrigation, requested assistance from the United Nations Development Program. A team was sent into the area in March 1966 to survey the problem. A request for funds was submitted and approved. The purpose of the -year project is to assist the government of India in undertaking a development program in the Chambal area as a basis for the formulation of a master plan for efficient use and management of soil and water for irrigated agriculture in Rajasthan.

Aquatic Weed Problem Areas

The Chambal River has, for many miles, a wide rocky bed which flows through narrow gorges. The water is very clear and is almost salt-free. Even during the monsoon season the Grandi Sagar Dam traps most of the silt in the river. Much of the area drained by this river in the Grandi Sagar has long been eroded away to a rocky bed. Aquatic plants will grow at very low light intensities, and in the clear waters of the Chambal canal system they can grow at great depths.

When constructing the right main canal, engineers used a series of farm tanks (old water storage ponds) as part of the canal. By using these tanks they saved about $400,000, and in addition the tanks were available for water storage in very dry seasons. These tanks have served as settling basins for any silt that might be in the canal waters and are a direct source of aquatic weed infestations for the newly constructed canal system. The use of these tanks by people and animals quickly spread the aquatic weeds through the entire main canal system.

The problem of aquatic weeds in the distribution canals and farm laterals is even more pronounced. Many of these canals are so badly choked with aquatic weeds that there is little flow of water. Some of the laterals have overflowed their banks because of flow restrictions. Drainage channels are very important in the Chambal irrigation scheme. New channels that are being constructed are choked with weeds soon after completion. The efficiency of older canals has been drastically reduced by emergent-type aquatic vegetation. The various aquatic plants that are considered to be a problem in the canals and the river are listed in Table 8.1.

Submersed aquatic weeds are causing the greatest problems in the canals. In this group the pondweeds (*Potamogeton spp.*) are dominant. *P. perfoliatus* comprises 75 to 80% of the submersed vegetation. The elodea reported in the system is believed to have been *Hydrilla verticillata.* Cattails (*Typha spp.*) are the major problem in the drainage channels.

Aquatic plants in the Chambal River were restricted to pools outside the main channel. The rocky bottom and fast flow prevented the establishment of aquatic plants. However, just below Kota Barrage, water hyacinth has become well established. All waters in the Chambal irrigation scheme are used by the local people for drinking, bathing and swimming, domestic animal consumption, and irrigation.

TABLE 8.1

List of Aquatic Plants Presenting Problems in Chambal Command Area (Plants Listed in Order of Importance in Each System)[a]

Common name	Scientific name
Large canals	
Redhead grass (S)	*Potamogeton perfoliatus* L.
Sago pondweed (S)	*Potamogeton pectinatus* L.
Tape grass (S)	*Vallisneria spiralis* L.
Marine naiad (S)	*Najas minor* Allioni
Hydrilla (S)	*Hydrilla verticillata* Royle
Minor canals	
Redhead grass (S)	*Potamogeton perfoliatus* L.
Sago pondweed (S)	*Potamogeton pectinatus* L.
Indian pondweed (S)	*Potamogeton indicus* L.
Hydrilla (S)	*Hydrilla verticillata* Royle
Marine naiad (S)	*Najas minor* Allioni
Drainage canals	
Narrowleaf cattail (E)	*Typha angustifalia* L.
Chambal River	
Water hyacinth (F)	*Eichhornia crassipes* (Mart.) Solms

[a]Classification based on growth habit, i.e., B, bank; E, emersed; F, floating; S, submersed.

Economic Considerations

The immensity of the aquatic weed problem has not been fully recognized by the local agencies. Local agencies say their requirements for irrigation are being supplied. Unless a value can be placed on the increased water movement through the canal system which would be brought about by proper weed control practices, it is very difficult to justify additional expenditures on aquatic weed control. The monetary value of increased water flow must be determined.

The cost of present weed control methods can only be estimated. In large main canals, cost estimates range from $600 to $1200/km. It is questionable if even the estimates of these methods should be discussed in this report. Since hand labor is the method presently used, only the cost of one annual cleaning of large canals is available from local agencies. This cost considers only the cost of employing additional local labor and does not take into consideration the cost of no water movement through the canal system for a period of 6 to 8 weeks, use of people regularly employed by the district, equipment costs, or human suffering. Certain canals were cleaned more than once a year, but no cost estimates per canal are available. The cost of maintaining these canals

under present control methods must be determined to justify aquatic weed control.

We must assume that as the irrigation acreage increases there will be a need for additional water. To supply this additional water there must be an effective weed control program. Even with maximum water movement through the system, it is not likely that there will be sufficient water to irrigate the original projected acreage for the project. Planting two crops annually per year is becoming an accepted practice in the area. This means that additional water will be needed for acreage that is already irrigated.

Present Methods of Aquatic Weed Control

Present methods of aquatic weed control used in the area for submersed weeds are hand cleaning and drying. During periods of water flow people wade in the canals or stand on the banks and pull weeds out. This method is used primarily in small canals with an average depth of 1 m. Larger canals are drained and the dry weeds are removed and piled on the bank by hand labor. The large right and left main canals are drained annually. The weeds are removed by hand, allowed to dry in the sun, and then piled and burned. Many areas of the larger canals never dry during the 6-week drawdown to the point that the weeds can be removed by hand. It requires a huge supply of labor to clean the main canals during water drawdown. The collection and transportation of labor to the working area becomes a problem. Those who are unskilled must be taught how to do the job.

Drainage channels are also cleaned by hand labor. The cattails are cut by hoe and are piled on the canal bank. Channels are usually reshaped during the cleaning. People living in the area often use the cut plants for binding up grain during harvest and for construction of thatched huts.

Authorities in the area have concluded that hand cleaning is not successful. It is a temporary control and in 2 to 3 months the small canals must again be drawn down when the weeds become a serious problem. However, the large canals can only be drained during the one period. The draining of canals more than once during the year causes crops to suffer from lack of water.

Project Operation

The maintenance budget for the irrigation department must be increased rapidly if aquatic weeds are to be controlled even at the present level. To increase control efficiency will require tremendous budget increases. Cost of maintenance may have to be absorbed by the farmer through increased irrigation charges. Financing the weed control organization can only be solved by the Chambal Command Area and the Rajasthan government. Immediate action should be taken on this matter if an efficient weed control organization is to be developed.

The urgency and vastness of the aquatic weed problem has not been fully recognized by the Chambal Command Area. Reduction in flow will be an additional 20% within the next 2 years.* This will mean a 75% reduction in the main canal flow rate in January through March. The reason for this increased weed growth is the aging of canals, extensive dispersal of weed seeds, and the increasing fertilization rates by farmers. The latter factor greatly increases the fertility of water in the canal system. Increased fertility will be further magnified by the double cropping of the farms.

Eradication of weeds in the canal system is not possible with present known methods of control. One can only hope to manage the weeds in the systems. A separate weed control organization was recommended by a specialist from the Food and Agriculture Organization of the United Nations (FAO) in 1966. This department still has not been formed by the state government. This weed control department must be organized, and must be staffed, equipped, trained, financed, and operated year round. Until this is done there is no hope of managing the aquatic weeds in the irrigation scheme.

Aquatic Weed Control Program

The aquatic weed control program in the Chambal Command Area was evaluated from the standpoint of its objectives as provided at the origination of the project. The project was to employ the services of a specialist who had experience with aquatic weeds in large flowing canals and streams. The inability of the project to obtain the services of such a specialist has seriously affected the program.

Fisheries Development

The three reservoirs of the Chambal valley development contain a water area of 563 km² at full pool. Grandi Sagar is located in Madhya Pradesh. Rana Pratrap Sagar and Jawahar Sagar (Kota Dam) are in Rajasthan. These reservoirs have

been constructed for the purpose of storing water collected in the monsoon season for irrigation and hydroelectric generation. No consideration was given to the value of these reservoirs as fresh water fisheries in the original planning of the project.

People of the command area are primarily vegetarians. It is against their religious beliefs to eat fish or other meats. However, many of the people in the villages and farm areas do eat fish. There has been a gradual migration of these people into the command area with a greater demand for fish. All fish that are being caught in the area are being consumed in the local market. Transportation of fish harvested in the area would not be a problem. Ice is not available in Kota. Because of the ice shortage all fish now caught in the area are sold in the local market very quickly.

Little information was available in Kota on the fishery in the Chambal Command Area. Fishermen were seen using both hook and lines and netting in the Chambal River. Fish were observed in both the river and canal system. No harvesting of fish was seen in the canal system even though fish were observed. Engineers of the Kota District stated that fish were collected in great numbers by some of the people when the canals were drained for weed cleaning. Fish move out of the Kota Barrage into the canal system and are trapped there when the canal gates are closed for draining.

The Rajasthan government has a fish biologist, Mr. A. Bhargw, stationed in Kota. A meeting was arranged with him in Kota, and he has supplied much of the information on fisheries of the area. The biologist feels there is a great need for the development of the fisheries in the Chambal Command area. Local fisherman are not able to meet the market demands for fish except during the canal drawdown. The Rajasthan government has become interested in stocking the reservoirs with fish but lack of funds has made progress very slow.

Fish culture ponds are under construction at Kethum and below Rana Pratrap Sagar Dam. Twelve ponds have been completed at Kethum, which is approximately 20 km from Kota along the right main canal. This will be all the ponds built in this area at the present time. The water supply for the ponds is the right main canal which is dry from April 15 to June 15 each year.

The possible use of fish as biocontrol agents for control of aquatic weeds has been discussed by many officials of Rajasthan. The grass carp (*Cteno-*

*by 1970.

pharyngodon idella) has been suggested as the fish most likely to be successful. The team that visited the Chambal Command Area in 1966 recommended that the fish be evaluated as a control agent for aquatic weeds. The grass carp would also serve as an additional source of food.

Manual methods of weed control will eventually become too expensive in the Chambal command area. The most economical method of control for any type of pest is the use of biocontrol agents. No biocontrol agent has been evaluated sufficiently to recommend its use in this vast water resource. It is to the advantage of all FAO nations that biocontrol with fish be given a careful evaluation. Biocontrol would be an additional step in the eventual management of the aquatic weed problem in the Chambal Command Area. There is a need for assistance in developing the fishery potential in this area. The market for fresh fish could be developed even with the disadvantage of religious beliefs, and the waters of this area should be highly productive. Development of fisheries would serve as an additional source of income for the project plus development of biocontrol for aquatic weeds in the Chambal canals and reservoirs.

EFFICACY OF THE WHITE AMUR

Research studies on biological control of aquatic weeds in India have been conducted at the Central Inland Fisheries Station, Cuttack, Onissa, India. The white amur herbivorous fish have been found to readily eat and control dense infestations of *Hydrilla, Najas, Ceratophyllum, Wolffia, Lemna,* and *Spirodela.* The fish also eat *Potamogeton pectinatus, Halophila ovata, Nitela, Spirogyra,* and *Pithophora,* but do not feed actively on *Eichhornia, Pistia, Nymphoides,* or *Aymphaea.* Information on the effective rate and size of the white amur indicates that, in *predator-free* waters, fingerlings of an average of 62 g were satisfactory for stocking. Where predators are present, the fish must be large enough to escape predation for satisfactory results. In preliminary tests, ponds choked with *Hydrilla* could be cleared by the fish at a stocking rate of 300 to 375 fish per hectare, in about a month's time. In general it is estimated that a stocking rate of 30 fish per hectare for a total weight of 35 kg would ensure satisfactory control of aquatic weeds.

The grass carp (*Ctenopharyngodon idella*) is one of the Chinese carp now being widely cultured. The species is a fast-growing one and feeds avidly on aquatic weeds, thereby utilizing what is considered obnoxious in fishery waters. In addition, the grass carp, in combination stocking with other cultivable varieties of fish, is reported to contribute indirectly towards their food by way of consuming the water weeds and in a way thereby green-manuring the ponds. The combination of all these qualities has made this vegetarian fish one of the most popular among the cultivable species. The observations made in China, Japan, Taiwan, Israel, Thailand, Malaya, India, Pakistan, the U.S.S.R., the U.S., and recently in the United Kingdom, recommend the fish.[1,2,13,22,36,37]

In the tropical countries, particularly in India, the "weed problem" in fishery waters is acute. Therefore, to supplement the other known methods of weed control and to achieve increased fish production, experimental consignments of grass carp were obtained from Hong Kong during December 1959 and January 1962, and the fingerlings were introduced at the Experimental Fish Farm, Killa, Cuttack. The preliminary observations made suggested the desirability of using the fish for aquatic weed control. Also, the possibility of breeding the fish in ponds by administration of fish pituitary hormones was demonstrated in 1962, and later large-scale production of its seed was achieved in 1964, enabling its distribution to almost all the states in the country for further observations in different regions. With an adequate stock of grass carp thus built up at the farm, a series of experiments were conducted in the field as well as at Cuttack Fish Farm to study and assess the efficacy of the fish in utilizing and controlling various water weeds.

Field experiments were conducted in protected ponds infested with aquatic weeds in Killa and Chaudwar fish farms (about 20 km from Cuttack) and in a few private ponds located in Cuttack proper. The density and species composition of the weed infestation were estimated statistically wherever possible, by using a 1-m^2 or 0.5-m quadrat frame employing the "systematic sampling" method. In situations where the infestation was rather scattered, sampling of the weed density was done only in the areas infested after assessing by "eye estimation" the percentage of coverage e.g., 25%, 50%, 75%, etc. In the case of patchy regenerating infestations no detailed sampling was considered necessary, but relevant field notes were made regarding the nature of infestation. Th observations on weed species not naturally growing in the protected ponds were made by introducing them in known quantities from other localities. As far as possible, control sets of weeds were maintained in order to form an idea about the natural changes taking place in the weeds (either growing naturally or introduced from outside). This was done by fixing up a 1-m^2 frame in a corner of the pond enclosing a proportionate sample of the healthy weed present outside the frame. Later, grass carps of known size and number were introduced in the ponds and regular observations on weed clearance were made. When the weeds under observation were fully cleared, the ponds were netted and final sampling on the survival and size of the fish was done (see Table 8.2).

Hydrilla verticillata Royle — Observations on control and consumption of *Hydrilla*, one of the most noxious water weeds, by grass carp were made by introducing the fish in ponds having a natural growth of the weed or by introducing the weed from outside. In one 0.08-ha pond, 450 kg of fresh *Hydrilla* was introduced and later 97 grass carp (average length 452 mm, average weight 955 g) were released which consumed all the weed within 8 days, and again 425 kg of the weed was

introduced which was consumed within 4 days. The fish, at the end of the experiment, averaged 1070 g.

Najas indica L. — This is another obnoxious weed having a very high aggressive capacity. The following observations were made in the field on its control and utilization. Two 0.08-ha ponds in the Killa fish farm, which were earlier dewatered and deepened, became thickly choked during the early monsoon season with *Najas indica,* and also with small scattered patches of *Nechamandra alternifolia, Nymphoides cristatum,* and *Marsilea quadrifolia.* In one pond a few *Vallisneria spiralis* clumps were also present. The weed density in the two ponds was sampled and was assessed to be 861 and 1103 kg, respectively. Control for *Najas* was maintained. Later 100 grass carp (average length 210 mm, average weight 94 g) were introduced in each pond. Within 6 weeks both the ponds were completely cleared of *Najas.* However, it was noted that in the absence of *Najas, Nechamandra* started spreading slowly, as did *Nymphoides.* Both these weeds, which were avoided by the grass carp, were manually removed at the time of final sampling. The fish in the two ponds measured an average of 341 mm and on an average weighed 470 and 474 g, respectively. Development of slight

TABLE 8.2

Effective Rate of Stocking of Grass Carp to Control Water Weeds

Weed species	Average wt (gm)	Stocking (fish/ha)	Weeds (tons/ha)	Clearance (days)
Hydrilla verticillata	995	1210	11	10
Najas indica	94	1250	10.8	41
	94	1250	13.8	41
	789	1667	19.0	14
Ceratophyllum demersum	2640	400	5.7	5
	616	1250	8.5	10
	830	1250	5.7	6
	623	1250	5.7	6
	974	250	37.2	49
Ultricularia stellaris	948	725	3.1	9
Spirodela polyrhiza	474	1250	6.5	20
Lemna trisulca	124	1000	1.7	11
	100	2000	3.6	9
Salvinia cucullata	958	1190	3.1	17

From Hickling, C. F., *J. Zool.,* 148, 408, 1966. With permission.

algal bloom was noticed in the ponds afterwards. The weeds inside the control remained healthy.

Ceratophyllum demersum L. — Observations on this submerged, uprooted, abundantly growing weed were also made. In an 0.08-ha pond 680 kg of *Ceratophyllum* was introduced and 100 grass carp (average weight 816 g) were stocked. Within 10 days all the weed was cleared and the fish finally weighed 623 g on an average. Once again in the same pond, 454 kg of *Ceratophyllum* mixed with some *Azolla* was introduced, and the fish took 6 days to consume the weeds. The initial and final average weights of the fish were 623 and 748 g, respectively.

Spirodela polyrhiza L. — Two 0.08-ha rearing ponds located side by side (which had been infested by a uniformly thick growth of *Najas* and had been simultaneously cleared by grass carp) were selected for observations on *Spirodela.* In each pond, 520 kg of healthy *Spirodela* mixed with some *Wolffia* was introduced; later, 100 grass carp were stocked in each pond (average length 341 mm) weighing on an average 474 and 470 g, respectively. Control for *Spirodela* was maintained.

Lemna trisulca L. — One 0.1-ha pond developed a thick cover of *Lemna* mixed with stray fronds of *Spirodela* and *Wolffia.* The total quantity of the weed was assessed by systematic sampling at 170 kg for the entire pond. Control for the weed was maintained and 100 grass carp (average length 211 mm, average weight 124 g) were introduced. The fish were observed feeding avidly on the duckweed, and the feeding activity was more vigorous during the morning hours. Within 11 days the pond was completely cleared of all the weeds, and on final sampling all the fish were recovered weighing on an average 145 g (average length 218 mm). The weed inside the control remained healthy and showed some increase in weight during the period of observation.

Utricularia stellaris L. — This is another submerged, unrooted weed, and its utilization by grass carp was also studied. Fresh *Utricularia* weighing 45 kg was introduced in one nursery pond in which 29 adult grass carp had been stocked. Within 5 days all the weed was consumed. Again in the same pond 125 kg of *Utricularia* mixed with some *Myriophyllum tuberculatum* was introduced

after sampling the grass carp present in the pond, which on an average weighed 948 g. Control for the weeds was maintained. In about 9 days the weeds were consumed, those inside the control remaining healthy. The fish finally averaged 975 g.

Salvinia cucullata Rox ex Bory — Fresh fronds of *Salvinia* weighing 250 kg, along with stray plants of *Trapa natans,* were introduced in one 0.08-ha pond, and control for the weeds was maintained. In the pond, 95 grass carp (average weight 958 g) were stocked. Within 17 days all the *Salvinia* and *Trapa* were consumed, though the weeds were not very actively fed upon. On final sampling the fish weighed 1 kg on an average. The weeds inside the control remained healthy.

PROJECT APPLICATION

It is surmised that the white amur (*Ctenopharyngodon idella*) is the most promising herbivorous fish for the biological control of a number of aquatic macrophytes, particularly submersed ones. The research results of various studies in different areas of the world and the U.S. support this assumption.[1,13,17,18,22,34,35] Although some information is available on the changes which will occur after introduction of the fish, many aspects concerning the influence of the white amur on the aquatic ecosystem have yet to be determined. To implement a new project, data already in the literature can be used to generalize the advantages and disadvantages of using the white amur for aquatic plant control.

Although biological control of unwanted plant growth appears to have many advantages over chemical and mechanical methods, biological control also has the potential for some disadvantages, particularly if the introduced species does not perform under natural conditions as it did under controlled conditions. There is no way of knowing exactly what effect the white amur will have on a large reservoir or canal system, or the possible changes which might occur when the white amur is used for aquatic plant management. It can be estimated from a review of established facts, and from general observations in the field, that this method should have general application in the different irrigation schemes.

REFERENCES – PART II

1. Avault, J. W., Jr., Preliminary studies with grass carp for aquatic weed control, *Prog. Fish Cult.,* 27, 207, 1965.
2. Barrington, E. J. W., The alimentary canal and digestion, in *The Physiology of Fishes,* Brown, M. E., Ed., Academic Press, New York, 1957, 109.
3. Bennett, F. D., Investigations on the insects attacking the aquatic ferns *Salvinia* spp. in Trinidad and Northern South America, *Proc. South. Weed Conf.,* 19, 497, 1966.
4. Bennett, F. D., Notes on the possibility of biological control of the water hyacinth, *Eichhornia crassipes, PANS Pest Artic. News Summ.,* 13, 305, 1967.
5. Bennett, F. D., Insects and mites as potential controlling agents of water hyacinth, *Eichhornia crassipes* (Mart.) Solms., *Proc. 9th Br. Weed Control Conf.,* 2, 832, 1968.
6. Anon., Pa Mong Project Interim Report, U.S. Bureau of Reclamation, U.S. Dept. of the Interior, Washington, D.C., 1969.
7. Chaffee, F. H. et al., Area Handbook for the Philippines (Department of the Army Pamphlet 550-72), U.S. Government Printing Office, Washington, D.C., 1969.
8. Cross, D. G., Aquatic weed control using grass carp, *J. Fish Biol.,* 1, 27, 1969.
9. De Datta, S. K., Lasscina, R. Q., and Seaman, D. E., Phenoxy acid herbicides for control of barnyard grass in transplanted rice, *Weed Sci.,* 19, 203, 1971.
10. Gangstad, E. O., Seaman, D. E., and Nelson, M. L., Potential growth of aquatic plants of the Lower Mekong River Basin, Laos-Thailand, *Hyacinth Control J.,* 10, 4, 1972.
11. Gangstad, E. O., Potential growth of aquatic plants in the Republic of the Philippines and projected methods of control, *J. Aquatic Plant Management,* 14, 10, 1976.
12. George, M. J., Studies on infestation of *Pistia stratiotes* Linn. by the caterpillar of *Manangana pectinicornis* Hamps., a Noctuid moth, and its effects on *Mansonioides* breeding, *Indian J. Malariol.,* 17(2-3), 149, 1963.
13. Hickling, C. F., On the feeding process of the white amur, *Ctenopharyngodon idella, J. Zool.,* 148, 408, 1966.
14. Holm, L., Weed problems in developing countries, *Weed Sci.,* 17, 113, 1969.
15. Holm, L. G., Weldon, L. W., and Blackburn, R. D., Aquatic weeds, *Science,* 166, 699, 1969.
16. Anon., House Document No. 389, Seventh and Final Report of the High Commissioner to the Philippines, U.S. Government Printing Office, Washington, D.C., 1947.
17. Kilgen, R. H. and Smitherman, R. O., Food habits of the white amur stocked in ponds alone and in combination with other species, *Prog. Fish Cult.,* 33, 123, 1971.
18. Lin, S. Y., Induced spawning of Chinese carps by pituitary injection in Taiwan, Chinese-American Joint Commission on Rural Reconstruction, Fisheries Serial 5, 1965.
19. Maddox, D. M. and Hambric, R. N., Use of alligator weed flea beetle in Texas; An exercise in environmental biology, *South. Weed Sci. Soc. Proc.,* 23, 283, 1970.
20. Maddox, D. M., Andres, L. A., Hennessey, R. D., Blackburn, R. D., and Spencer, N. R., Insects to control alligatorweed: An invader of aquatic ecosystems in the United States, *BioScience,* 21, 985, 1971.
21. Maheshwari, J. K., Alligator weed in Indian lakes, *Nature,* 206(4990), 1270, 1965.
22. Michewicz, J. E., Sutton, D. L., and Blackburn, R. D., Water quality of small enclosures stocked with white amur, *Hyacinth Control J.,* 10, 22, 1972.
23. Machenthun, K. M. and Ingram, W. M., Biologically Associated Problems in Freshwater Environments, Federal Water Pollution Control Administration, U.S. Dept. of the Interior, Washington, D.C., 1967.
24. NagRaj, T. R. and Ponnappa, K. M., Some interesting fungi of India, *Tech. Bull. Commonw. Inst. Biol. Control,* 9, 31, 1967.
25. Anon., *Principles of Plant and Animal Pest Control,* Vol. 2, *Weed Control,* National Academy of Sciences, Washington, D.C., 1968.
26. Pancho, J. V., Vega, M. R., and Plucknett, D. L., *Some Common Weeds of the Philippines,* Weed Science Society of the Philippines, Manila, Philippine Islands, 1949.
27. Rao, V. P., Possibilities of Biological Control of Aquatic Weeds in India, paper presented to the Annual Research Session of the Central Board of Irrigation and Power, Ahmedabad, India, 1969.
28. Raven, P. H., The Old World species of *Ludwigia* (including *Jussiaea*), with a synopsis of the genus (*Onagraceae*), *Reinwardtia,* 6(4), 327, 1963.
29. Sankaran, T., Srinath, D., and Krishna, K., Studies on *Gesonula punctifrons* Stal (Orthoptera: Acrididae: Cyrtacanthacridinae) attacking water hyacinth in India, *Entomophaga,* 11, 433, 1966.
30. Sankaran, T., Srinath, D., and Krishna, K., *Haltica caerulea* Olivier (Col.: Halticidae) as a possible agent of biological control of *Jussiaea repens* L., *Tech. Bull. Commonw. Inst. Biol. Control,* 8, 117, 1967.
31. Sankaran, T. and Krishna, K., The biology of *Nanophyles* sp. (Col.: Curculionidae) infesting *Jussieua repens* in India, *Bull. Entomol. Res.,* 57, 337, 1967.
32. Strivastava, J. G., Some tropical American and African weeds that have invaded the State of Bihar, *J. Indian Bot. Soc.,* 43, 102, 1964.
33. Stephans, J. C., Blackburn, R. D., Seaman, D. E., and Weldon, L. W., Flow retardance by channel weeds and their control, *ASCE J. Irrig. Drain. Div.,* 89, 31, 1963.

34. Stevenson, J. H., Observations on grass carp in Arkansas, *Prog. Fish Cult.*, 27, 66, 1965.
35. Sutton, D. L., Utilization of hydrilla by the white amur, *Hyacinth Control J.*, 12, 66, 1974.
36. Swingle, H. S., Control of pondweeds by use of herbivorous fishes, *Proc. South. Weed Conf.*, 10, 11, 1957.
37. Tang, Y. A., Hwang, Y. W., and Lin, C. K., Reproduction of the Chinese carps, *Ctenopharyngodon idellus* and *Hypophthalmichthys molitrix* in a reservoir in Taiwan, *Jpn. J. Ichthyol.*, 13(1/2), 1, 1960.
38. Anon., Central Luzon Basin, USAID/Philippines, U.S. Bureau of Reclamation, 1966.
39. Anon., Upper Pampanga River Project, USAID/Philippines, U.S. Bureau of Reclamation, 1966.
40. Anon., Cagayan River Basin, USAID/Philippines, U.S.Bureau of Reclamation, 1966.
41. Anon., Ilog-Hilabangan River Basin, Negros Island, USAID/Philippines, U.S. Bureau of Reclamation, 1966.
42. Anon., Agusan River Basin, Mindanao Island, USAID/Philippines, U.S. Bureau of Reclamation, 1966.
43. Anon., Cotabato River Basin, Mindanao Island, USAID/Philippines, U.S. Bureau of Reclamation, 1966.
44. Anon., Bicol River Basin, USAID/Philippines, U.S. Bureau of Reclamation, 1967.

Part III
Research for Control of Alligator Weed

Chapter 9

INTRODUCTION TO ALLIGATOR WEED*

The primary weed of consideration in this part is alligator weed (*Alternanthera philoxeroides* [Mart.] Griseb.). A herbarium sheet deposited in the herbarium of the United States National Arboretum indicates that Major William Rossel, U.S. Army Corps of Engineers, was probably the first to recognize this weed as a menace. The specimen was deposited in the National Arboretum Herbarium under the following label:

Telanthera philoxeroides Mog. "Duck Grass." A serious obstruction to navigation and drainage in the Mobile River and its tributaries. Mobile, Alabama. Collector Major William Rossel, U.S.A. Eng. Corps, July 9, 1901. U.S. Nat. Arb. Herb. No. 44446.

The question might well be asked why was this plant not more generally recognized as a hazard. It will be remembered that the water hyacinth was reputedly introduced in the United States in 1884 at the New Orleans Cotton Exposition. Numerous collections were made of this plant in the late 1890s and in the early 1900s from such widely separated points as Texas and Florida, indicating that the plant had by the early part of this century well established itself along the Gulf coastal section of the United States. It is well known that if the water hyacinth in any given season gets a head start on the alligator weed, it will outgrow the alligator weed and to a great extent shade out and retard the alligator weed. Therefore, the water hyacinth was regarded as the primary pest in the minds of most people, and alligator weed was to some extent disregarded. It was only after the systematic removal of the water hyacinth with 2,4-D that the alligator weed returned as the potential threat pointed out by Major Rossel in 1901.

AQUATIC PLANT USE

The alligator weed has long been recognized as noxious growth in many agricultural lands in the southeastern part of the United States and since this terrestrial plant has adapted itself to an aquatic environment, it has been classified as an aquatic weed. One exception to this might be indicated in the attitude of certain cattlemen in the southwestern parishes of Louisiana, who to some extent depend on the alligator weed for winter feeding of their cattle; however, in the overall picture, the alligator weed is more undesirable than desirable. Other aquatic plants present similar problems. The duckweed, also regarded as a common aquatic pest, does have some value as a wildlife food plant. Moreover, studies made on a large lake used by the Central Louisiana Electric Company for cooling waters for power condensers revealed that during the summer when maximum cooling effects are desired, the floating duckweed produced a shield against solar radiation, thus preventing the lake — acting as a black body — from absorbing a certain percent of solar radiation. In this instance the duckweed proved to be beneficial. This does not say, however, that duckweed is always beneficial, nor is it in fact beneficial in the general area of the lake, but only for the specific purpose of keeping the lake waters cool in summer.

AQUATIC PLANT CONTROL

Over the many centuries that man has changed from nomadic to settled agrarian life, there has developed simultaneously the need to eliminate noxious vegetation from cultivated lands. In the most primitive beginnings of civilization, man noticed that undesirable plants interfered with his attempts to grow useful plants and, in an attempt to give the useful plants a better chance to survive, he resorted to a process of hand weeding. As civilization advanced, man devised crude instruments to assist him in removing noxious vegetation from his desirable crops. Very primitive hoes and other devices were used in the fertile valleys of the Euphrates and the Nile Rivers. In time, these hand-manipulated tools gave way to devices that could be pulled or drawn by animals. The materials of construction of the weeding implements gradually changed from wood and stone to bronze and steel as man advanced to the Iron Age. At some point in his civilization, man discovered that such compounds as salt would sterilize the soil and prevent the growth of certain plants; thus, salt was a primitive chemical method of controlling weeds.

*Abridged and updated from unpublished report, "Control of Alligatorweed and Other Aquatic Plants," University of Southwestern Louisiana, Lafayette, Louisiana, 1967.

In the Middle Ages it was noted that other chemical compounds had desirable herbicidal properties. Arsenic compounds, for example, were used to control plant growth. In the early part of the 20th century, it was noted that other organic compounds had a herbicidal effect on some plants. Today, most research effort is directed toward the use of organic compounds as herbicides.

PHYTOTOXICITY

During the period from 1930 to 1950, certain chemical compounds were found to affect the metabolism of certain plants and promoted or retarded growth. In some cases the promoting or retarding was merely a question of the quantity or concentration of compounds that could be absorbed into the plant. The chemical compound known as (2,4,-dichlorophenoxy)acetic acid was observed to have unusual and unique properties as a herbicide. Claims have also been made that when this compound is used in very low concentration, it is possible to increase crop yields or produce parthenocarpsis. This parent compound, that is, (2,4-dichlorophenoxy)acetic acid, commonly known as 2,4-D, has many useful derivatives. In general, phytotoxicity of the derivatives of this compound are related to the toxic properties of the parent acid. The derivative compounds are sometimes more desirable than the parent acid because of their physical properties; that is to say, they might have a greater solubility in water or greater solubility in oil which would allow them to be disbursed more readily in a two-phase emulsified system, or they might have such properties as low volatility which prevent their vaporization and disbursal into unwanted areas.

HERBICIDE DEVELOPMENT

There has been a tendency on the part of a great many research workers to accept herbicides manufactured by some particular company and to adapt these as they are to a particular aquatic plant problem. These adaptations usually consist of altering the amount per acre or number of applications or some such empirical technique. Although many have been successfully used for this special purpose, none have been as spectacular in their results as the phenoxy compounds. One of the common methods used by the chemical companies in determining the potential use of a

compound as an aquatic herbicide has been to simply take a backlog of related chemical compounds on the shelf and try them out.

ECONOMIC ADVANTAGES
OF CONTROL

Economic advantages of the control of aquatic vegetation are not always clearly evident. In most cases aquatic vegetation is not affecting the production of an economic crop. Usually the major effects are on inland fishing and, in particular, on sport fishing, recreation, and/or navigation. Because of this fact it is almost essential that federal and other governmental agencies take over the problem of aquatic weed control. Moreover, bodies of water involved in an infestation are not usually the sole property of an individual but of many individuals and/or the property of a governmental body. Thus the problem often becomes an area problem rather than a highly localized problem. It may, however, present a very real local problem to the homeowner or boat owner on a lake or canal.

It has been estimated that the annual loss from weeds in the United States approximates $5 billion. A 5-year study (1946 to 1950) made in Louisiana indicated that the probable annual loss in Louisiana due to aquatic vegetation was approximately $35 million. Water hyacinth was the most prevalent aquatic weed observed in this study. This particular pest interfered with recreation, fish, and wildlife, and brought about abnormal streamflow conditions that resulted in the flooding of agricultural lands and is currently a problem.

The Corps of Engineers developed several mechanical devices to control the water hyacinth. Draglines were used in certain areas to pile the plant on shore. Later devices were used in which moving belts either conveyed the water hyacinth to the shore where it was allowed to dehydrate and die, or moving belts lifted the plant bodily from the water and dumped the plant material between heavy rollers, thus crushing the plant and returning the pulp to the water in which the destroying device operated. Barge boats with a high-speed rotating shaft of saw blades at approximately 2-cm intervals were also used. These saw operated near the surface of the water and shredded or cut the plants into many small pieces This was a satisfactory method for opening streams entirely clogged with the water hyacinths, but when the water hyacinth is removed from a waterway, it is often succeeded by alligator weed, if that growth was present before the removal operations. Depending upon conditions, a stream may be infested alternately by alligator weed and water hyacinth as removal operations progress.[12,13]

On the basis of extensive experience in controlling the plant in thoroughly infested waters, and on the basis of spot checks on private operators who occasionally operate vessels in such waters, it is concluded that, if the present control operation was stopped, the cost of navigation for all classes of traffic in waterways choked with aquatic growth would be at least doubled as long as the traffic continued to struggle through. On the basis of the 2,600,069,095 t-mi for these waterways, the operating loss for navigation would be $6,500,000. For those secondary and other streams for traffic carried largely by the small craft of individual owners, it is estimated that the total traffic is 50,000,000 t-mi per year at a normal operation cost of $0.01 per tonne-mile. The actual present losses in this unreported traffic due to these aquatic plants are estimated to be 50% of the operating cost, or $250,000 per year. The total annual losses would be $6,750,000, and this sum may be taken as the theoretical total navigation benefit from an effective control program.

Perhaps the outstanding impact of water hyacinth and alligator weed on fish and wildlife is that a blanket of these growths on the surface of any open water can destroy all fish and wildlife, by stopping the exchange processes that go on constantly between the atmosphere and any body of water and which are essential to the whole cycle of aquatic organic life. Thus open lakes, ponds and reclamation lakes, marsh swales, slow streams, canals and borrow pits, flood bottoms, and fresh and salt marshes are all subject to damage by these aquatic plants and may be threatened ultimately with a nearly complete loss of usefulness.

Aquatic environments are constantly changing; when left undisturbed for long periods, their vegetation follows a successional pattern that culminates in a grassland climax. The coastal marshes are most productive of wildlife in their lower successional stages. In climax, the wildlife food plants of the latter stages are obliterated. Prolonged floods, hurricane tides, and severe droughts with attendant fires all serve to remove worthless climax plants. Valuable wildlife food plants are then quick to recover, and although wildlife may have suffered during the floods or droughts, it is benefited in the long run by an improvement of its habitat. But the same floods and storm tides today also distribute the water hyacinth and alligator weed and, as a consequence, the natural factors that should bring improvements to habitats are actually spreading the very plants that destroy them.

Chapter 10
CHEMICAL LABORATORY STUDIES*

PLANT CHARACTERISTICS

Preliminary studies on the chemical control of alligator weed and water hyacinth with the phenoxy herbicides were made in 1950 under a research contract of the U.S. Army Corps of Engineers (New Orleans District), Boyce Thompson Institute for Plant Research, and Tulane University.[4] Since that time, phenoxy herbicides have been widely used for the control of water hyacinth in Florida, Louisiana, and other states of the Gulf Coast. Although the use of phenoxy herbicides for control of water hyacinth has been very successful, treatment of alligator weed has been less satisfactory, because of the high levels of application required and the frequent retreatment necessary to keep the plant under control. To improve this method of control, research studies were initiated in 1959 under a research contract with the New Orleans District of the U.S. Army Corps of Engineers, the Office of the Chief of Engineers, and the University of Southwestern Louisiana.

It is common practice to apply the chlorophenoxy herbicides to the foliage of plants. Foliar application involves certain external factors which determine the ability of the herbicide to give either good or poor penetration. One of the most commonly observed factors is the addition of surfactants to the aqueous solutions being sprayed. This gives better penetration because the liquid film covers a greater leaf area and actually penetrates the outer wax layer and causes the liquid spray to enter through the leaf wax more easily. It has also been observed that the outer leaf surface is composed of a rather unreactive cuticle, which is composed of waxes, esters, and water-insoluble cyclic compounds. The water-soluble esters of the chlorophenoxy herbicides dissolve in this wax layer, or in the case of abrasions in the cuticle, with sufficient surfactant present penetrate this layer (see Figures 10.1 to 10.4).

Immediately below the cuticle are cell walls composed primarily of cellulose and pectins. This layer is easily penetrated with water-soluble compounds. If there are deep cracks in the cuticle, this layer is directly exposed to herbicidal spray, provided the surface tension of the spray solution is low enough to allow penetration. Another route of entry for the herbicide is through the stomata. The chlorophenoxy herbicides concentrate in the symplast and are moved to the phloem. These compounds are then moved, in the plant, to points of active growth (see Figures 10.5 to 10.8). In the alligator weed, it is difficult to obtain good absorption and herbicidal action with phenoxy-type herbicides after senescence begins because movement in the phloem is reduced. Under this situation, absorption through the root system is apt to provide a better mode of entry for the herbicide.

ORGANIC ACID INHIBITION

Chlorophenoxy herbicides undergo degradation within plants. Organic acids are degradation products of this metabolism. Accordingly, organic acids were studied for root and foliage absorption to observe possible synergistic effects. For this study, vigorous field shoots 60 cm long (30 cm above and 30 cm below the waterline) were placed in beakers under controlled light and air temperatures. Observations were made 2 and 4 weeks after treatment, and the degree of control in percent live tissue estimated. Data on the effects of 2,4-D and some organic acids are summarized in Table 10.1. Although disintegration of the tissue was significantly affected by the treatments, subsequent field tests did not improve herbicide action sufficiently to merit special formulation.

TRANSLOCATION OF 2,4-D

Basic studies were initiated with tracer chemicals to observe the translocation of 2,4-D in alligator weed. The tagged materials were applied according to methods used for tracer studies of 2,4-D.[4] Uniform plants were selected from a greenhouse nursery, nine plants for each treatment. Insofar as possible, the young plants were of uniform height, having the same number of nodes

*Research contribution of Dr. F. W. Zurburg, Director of Research, and S. L. Solymosy, Botanist, University of Southwestern Louisiana, Lafayette, Louisiana.

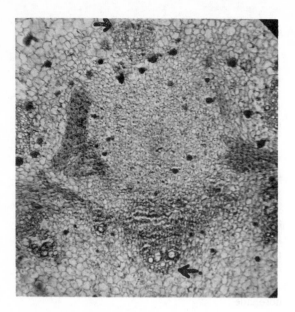

FIGURE 10.1. Cross section of a stem. As shown by the arrow, the phloem occurs only on the outer side of the xylem cylinder.

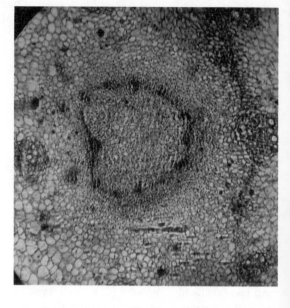

FIGURE 10.3. Cross section of bud, lower section. No direct connection is observed between the bud and the vascular tissue of the parent stem.

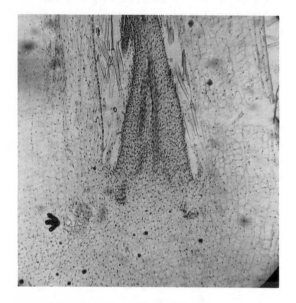

FIGURE 10.2. Longitudinal section of a node. As shown by the arrow, axillary buds arise exogenously in superficial tissues.

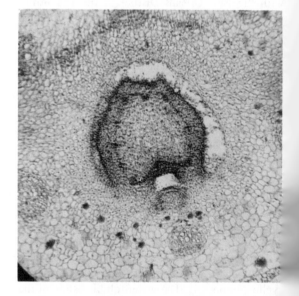

FIGURE 10.4. Cross section of bud, upper section. Each bud develops into a shoot and duplicates the vascular pattern found in the parent shoot apex.

and other plant parts. The rooted young plants were then moved into proper containers, treated in the laboratory with a tagged chemical, and taken back to the greenhouse. Concentrations of radiochemicals used in the following experiments were measured in parts per million.

In order to confine these chemicals, barrier were built of lanolin, greases, or paste, which wer known to be inactive and which could hold th application at the desired point. In the unique cas of root application, preventing moisture loss wa necessary to prevent drying of the roots. There

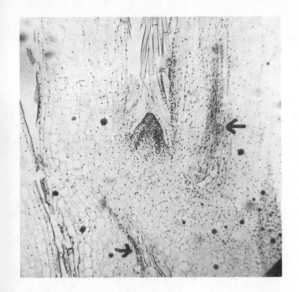

FIGURE 10.5. Longitudinal section of node, 2 kg acid equivalent per hectare formulation, 2 days after treatment. Some breakdown of tissue is observed. Only vascular tissue is affected.

FIGURE 10.7. Longitudinal section of node, 2 kg acid equivalent per hectare formulation, 8 days after treatment. Effect of the herbicide is not severe, allowing further translocation to other parts of the plant.

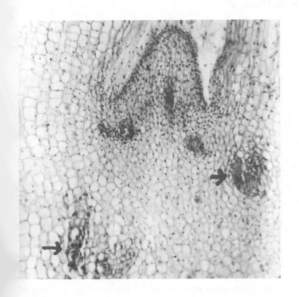

FIGURE 10.6. Longitudinal section of node, 4 kg equivalent per hectare formulation, 2 days after treatment. The base of the leaf trace shows some destruction of tissue.

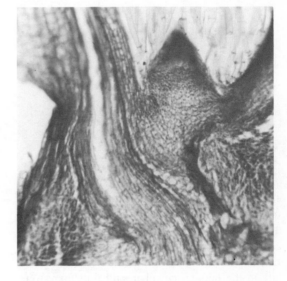

FIGURE 10.8. Longitudinal section of node, 4 kg acid equivalent per hectare formulation, 8 days after treatment. Effect of the herbicide is severe and extensive breakdown of the tissue has occurred, with the exception of the axillary bud, which may begin new growth.

ore, a device was built which would keep the lant in a saturated atmosphere and, at the same me, allow application of the tagged chemicals.

Counting equipment consisted of a Tracer® aboratory Versa/matic V scaler. The plant mate-

rial was exposed to a counting tube and an atmosphere of argon. The geometry and window absorption and other factors were checked against a standardized radioactive reference source of carbon-14. The actual disintegration rate compared with the standard gave it a total overall

TABLE 10.1

Herbicidal Activity of Combinations of 2,4-D and Organic Acids on Alligator Weed

Beaker number	Organic acid	Concentration (ppm)	Herbicide placement	Plant[a] response	% Control[b]
1a	Levulinic acid	500	Top	Epinasty	0
1b	Levulinic acid	500	Root	Epinasty	
1c	LLA + 2,4-D	500 + 50	Top	Collapse	100[c]
1d	LLA + 2,4-D	500 + 50	Root	Collapse	100[c]
2a	Methoxyacetic acid	500	Top	Chlorosis	0
2b	Methoxyacetic acid	500	Root	Chlorosis	0
2c	MOA + 2,4-D	500 + 50	Top	Collapse	100[c]
2d	MOA + 2,4-D	500 + 50	Root	Collapse	100[c]
3a	Oxalic acid	500	Top	Chlorosis	30[d]
3b	Oxalic acid	500	Root	Chlorosis	60[c]
3c	OA + 2,4-D	500 + 50	Top	Collapse	80[c]
3d	OA + 2,4-D	500 + 50	Root	Collapse	100[c]
4a	Diglycolic acid	10	Top	Collapse	20[e]
4b	Diglycolic acid	10	Root	Collapse	60[c]
4c	DGA + 2,4-D	10 + 50	Top	Collapse	100[c]
4d	DGA + 2,4-D	10 + 50	Root	Collapse	100[c]

Summary of Treatment Means

1a + 2a + 3a + 4a			Top		12 NS[e]
1b + 2b + 3b + 4b			Root		30
1c + 2c + 3c + 4c			Top		95[e]
1d + 2d + 3d + 4d			Root		100

[a]Plant response 2 weeks after treatment.
[b]Plant response 4 weeks after treatment.
[c]Highly significant response, P = 0.01.
[d]Significant response, P = 0.05.
[e]Nonsignificant.

efficiency of 21.4%. Little difficulty was encountered in the counting technique since the procedure was more or less standard. Even though these acids are not very volatile under normal conditions, sufficient amounts did evaporate to accumulate in the counter chamber, and thus there were background buildups. For this reason, background counts were always made once at the start and again at the end of the experiment when volatile compounds were used. The background was also checked intermediate to these two checks.

Since having samples of the same size would obviously be an advantage, a slicing device was regulated to cut sections from the plant approximately 0.25 mm in thickness. It was realized that the density of the material in the internodes would be much less than the density in the nodal section; therefore, a microbalance was used to determine the weight of the sample so that the eventual reading could be given in terms of disintegration per second (dps) per milligram of plant material.

The freezing technique was used to obtain the data presented immediately below. The herbicide used was 2,4-D at 1000 ppm dissolved in SX-5412, which is an 85% aromatic hydrocarbon solvent manufactured by Humble Oil Company. Fifty aliquots of this solution was applied to the two leaves attached to the third node. The herbicide was ^{14}C-labeled in the ring. Since treatments were applied only to the leaves attached to the third node, the herbicide must move from this node to the other plant parts. Consideration must be given not only to the fact that the plant is withdrawing herbicide from this node but also to the fact that this node in turn is continually supplied with additional 2,4-D from the surface of the leaves attached to this node. Therefore, this node may be looked upon as a reservoir which is capable of feeding 2,4-D to the rest of the plant as it derives additional 2,4-D from the leaves.

DISTRIBUTION OF RADIOACTIVE HERBICIDE

The data indicate that there is a rapid upward movement toward the growth point. Some downward movement is also noted. At the end of 12 hr the plant was in darkness and thus the distribution of the herbicide was more or less uniform, with the greatest concentration appearing in the third internode (the internode just below the point of application). At the end of 24 hr the plant was again in sunlight, with the upper points of the plant again receiving the major part of the herbicide. The lowest node (or fifth node to which the roots were attached) received very little herbicide. The movement upward was found to be greater than the movement downward, which explains in part why the top is readily killed by the herbicide and the rootstock or mat is not. The relative distribution at the end of 24 hr was found to be 3000 pg upward, 375 pg lateral, and 575 pg downward.

Chapter 11

BOTANICAL LABORATORY STUDIES*

HERBARIUM COLLECTION

Members of the research staff at the University of Southwestern Louisiana (U.S.L.) are engaged in collecting plant specimens for a herbarium. Alligator weed plants are filed in the University herbarium and are classified according to Correll and Correll[2] Alternanthera Philoxeroides (Mart.) Griseb. (*J. Wash. Acad. Sci.* 5:74, 1915). Synonyms are

● *Bucholzia philoxeroides,* Mart. (Nova Acta Acad. Leop. Carol. 13:315, 1826)
● *Telanthera philoxeroides,* Moq. (DC. Prodr. 13:362, 1849)
● *Thelanthera philoxeroides* obtusifolia, Moq. (DC. Prodr. 13:363, 1849)
● *Telanthera philoxeroides acutifolia,* Moq. (DC. Prodr. 13:363, 1849)
● *Telanthera philoxeroides phyllantha,* Seub. (Mart. Fl. Bras. 5:169, 1875)
● *Telanthera philoxeroides denticulata,* Seub. (Mart. Fl. Bras. 5:169, 1875)
● *Alternanthera philoxeroides,* Griseb. (Abh. Ges. Wiss. Goett. 24:36, 1879)
● Common name: Alligator weed

IDENTIFICATION AND MORPHOLOGY

Alligator weed was originally identified as *Achyranthes philoxeroides* (Mart.) Standley, Family: Ameranthaceae; Tribe: Gomphrenae. The tribe Gomphrenae differs from other tribes of the family by a solitary pendulous ovule, ascending radiculus, two-celled anthers dehiscing by longitudinal splits, hypogynous stamens and free perianth segments. This species differs from other species of other genera by pedunculate flower heads, sessile flowers, and glabrous sepals. The plant is a herbacious perennial, mostly fistulous, ascending in terrestrial and decumbent in aquatic habitat, branched (rarely simple), villous in the leaf axils and around the nodes, sparsely villous or glabrous elsewhere; leaves short-petioled, sometimes sessile, peduncles axillary and/or terminal, always simple, pilose, sometimes glabrous, spikes globose or elongate-globose; flowers white, apetalous, pentasepalous, subtended by short bractlets, sepals chartaceous, acute, more or less obscurely veined, often lacerate, serrulate toward the apex; anther tube composed of five anthers alternating with five staminodia, staminodia lacerate at apex; stigma entire, capitate, fruit a one-seeded utricula. Differences in measurements between the descriptions found in the existing literature and the collected plant material are shown in Table 11.1.

A. philoxeroides is a polygamous plant; male, female, and perfect flowers have been found on the same torus; 242 inflorescences with 5915 florets from Lafayette, St. Martin, Evangeline, and Cameron Parishes (all in Louisiana) have been examined. The results are shown in Table 11.2.

HABITAT CHARACTERISTICS

Alternantheria philoxeroides is a true amphibious plant; it can be found growing from almost arid conditions to swampy areas and open water. No differences in vigor or general appearance have been observed whether grown on land or in water. According to Bergman's law, the maximum size of the species is found in the optimal region; plants of *A. philoxeroides* with the largest stem diameter and the longest stem have been found among melangeophilous, and plants covering the largest area have been found among telmicolous and lacustrine, specimens.

On land, *A. philoxeroides* is found growing upright mostly around and close to shrubs, sometimes covering a good part of their surface using the structure as a support.

Growing in water the plant becomes decumbent. One specimen has been collected in slightly brackish water (University of Southwestern Louisiana Ornamental Horticulture Herbarium No. S042161126). The plants are anchored by their roots to the shore, or in shallow water to the bottom. In deeper water they are detached and form floating mats of considerable size. Plants have been collected and exsiccata deposited in the

Research contribution of Dr. S. L. Solymosy, Botanist, Botanical Laboratory, University of Southwestern Louisiana. Lafayette, Louisiana.

TABLE 11.1

Plant Measurements of Herbarium Specimens from Different Sources

		N. Am. Flora[a]	Small[b]	Muenscher[c]	U.S.L. check[d]
Leaves, cm	Length	3.5–10.0	3–11	–	2.0–14.4
	Width	0.5–2.0	–	–	0.2–2.0
Peduncle, cm		1.0–5.0	–	–	0.9–7.0
Bracts, mm	Length	¼ of sepal	–	–	1.8–2.5
	Width	–	–	–	0.2–0.3
Male flower					
Sepals, mm	Length	6	6	–	6.5–9.0
	Width	–	–	–	1.0–2.0
Stamen, mm	Filament	½ of sepal	–	3.0–3.5	
	Anther	–	–	–	0.8–1.0
Spike, cm	Length	1.4–1.7	–	–	1.1–2.0
	Width	–	–	–	1.3–1.8
Female flower					
Sepals, cm	Length	6	6	6	6.0–6.5
	Width	–	–	–	2.5–3.0
Utricle, mm	Length	–	–	–	4.0–5.0
	Width	–	–	–	5.0–6.0
Seed, mm	Diameter	–	–	–	0.9–1.5
Spike, cm	Length	1.4–1.7	–	–	0.7–1.2
	Width	–	–	–	1.2–1.4

[a]*North American Flora,* The New York, Botanical Garden, New York, 1917.
[b]*Southeastern Flora,* John Kunkel Small, New York, 1933.
[c]W. C. Muenscher, *Aquatic Plants of the U.S.,* Comstock, Ithaca, N.Y., 1944.
[d]*Aquatic and March Plants of Louisiana,* University of Southwestern Louisiana, Lafayette, La., 1969.

TABLE 11.2

Number of Spikes and Flower Type for Plants in the University of Southwestern Louisiana Ornamental Horticultural Herbarium

Parish	No. of spikes	Individual flowers			
		Male	Female	Perfect	Total
Lafayette	104	300	100	2200	2600
St. Martin	56	660	–	600	1260
Evangeline	52	300	300	695	1295
Cameron	30	210	185	365	760
Total	242	1470	585	3860	5915
Ratio		24%	9.8%	66.2%	100%

University of Southwestern Louisiana Ornamental Horticulture Herbarium as shown in Table 11.3.

SEXUAL REPRODUCTION

Very little can be reported on the sexual reproduction of *A. philoxeroides.* Experiments in the field and under laboratory conditions showed very few positive results; most findings were negative. *A. philoxeroides* is everblooming. In Louisiana the flowering begins in early spring, reaching a peak several times within a year, and ceases upon arrival of the first frost. In the Lafayette area the first blooming starts in March to April, reaching its peak in about 2 weeks; the next high is reached in early June, the following i

TABLE 11.3

Habitat Characteristics for Specimens in the USL Ornamental Horticultural Herbarium

Date	Location	Collection no.	Aquatic (A) or terrestrial (T)
6-10-60	Iberia Parish: New Iberia, sandy loam.	SO619611	T
6-21-61	Lafayette Parish: 2 mi west of Youngsville, roadside ditch.	SO621601	T
7-9-60	Lafayette Parish: 2 mi west of Broussard, pond.	SO709601	A
7-9-60	St. Martin Parish: Martin Lake.	SO709602	A
7-11-60	St. Martin Parish: 2 mi north of St. Martinville, silt.	SO711601	T
7-15-60	Lafayette Parish: Lafayette U.S.L. Horticulture Farm, under trees.	SO715601	T
7-20-60	Lafayette Parish: 2 mi west of Broussard, pond.	SO720601	A
7-20-60	Lafayette Parish: U.S.L. Dairy Farm, pasture.	SO720602	T
7-27-60	Lafayette Parish: 2 mi southeast of Lafayette.	SO727601	A
7-28-60	St. Martin Parish: St. Martinville, along Bayou Teche.	SO728601	A
7-30-60	Lafayette Parish: Lafayette on high banks of Coulee Mine, sandy loam.	SO730601	T
8-9-60	Vermilion Parish: 4 mi south of Perry, wet depression.	SO8096012	A
8-17-60	Sabine Parish: west of Sandell (Hodges experimental area), lake.	SO8176015	A
8-23-60	St. John The Baptist Parish: Lake Maurepas.	SO8230601	A
8-25-60	Cameron Parish: south of Cameron, in canals.	SO8256050	A
9-3-60	St. Martin Parish: Bayou Teche.	SO903601	A
9-4-60	Evangeline Parish: Chicot State Park, in lake.	SO904607	A
9-4-60	St. Martin Parish: Ward 5, in lake.	SO904601	A
10-4-60	Lafayette Parish: St. Martin Lake.	S1004606	A
10-7-60	Iberia Parish: Avery Island, sandy loam.	S1007604	T
4-21-61	Alabama, Baldwin County: about 4 mi east of Gulf Shores, slightly brackish lagoon.	SO42161126	A
4-15-65	St. Martin Parish: Martin Lake, in shallow water.	SO415651	A
12-10-64	Vermilion Parish: between forked Island and Pecan Island.	S1210641	T

mid or late July, and the last in late August; from this date until the first frost a constant but sporadic blooming takes place. Under laboratory conditions the plant blooms constantly with a low around December 22, though never reaching the same profusion as in the field.

SEXUAL ORGANS OF THE FLORET

The florets of the inflorescence develop from the bottom up. The anther tube envelopes the gynoecium forming a kind of protective cover. At anthesis the tube opens like a crown, at night, on cloudy days, or under dark shade, thus exposing the whole gynoecium.

The anther tube open, the stigma secretes a clear honeylike substance to show its readiness to accept the pollen. Once fertilized (or if not, usually after 48 hr) the movement of the anther tube ceases, the androecium stays open, the anthers begin to wilt and dry, and the utricles begin to swell, provided fertilization took place. This phenomenon leads to the belief that the pollination is done at night or during cloudy days by night-flying or heliofuge pollinators. Viable seeds were collected only in July (U.S.L. Ornamental Horticulture Herbarium, S0727601). Three seeds were dissected; two were planted in a petri dish using milled sphagnum as medium. They did not germinate. Further collecting trials were unsuccessful.

Hand pollination was tried under crude laboratory conditions on florets with open anther tubes and on some with closed tubes. The anther tubes were forced open with a dissecting needle and the pollen was applied on the stigma. On florets with expanded anther tubes the utricles swelled and kept swelling for 6 days, then shriveled and dried. On the ones with closed anther tubes, the utricles did not swell at all.

VEGETATIVE REPRODUCTION

The occurrence of the sexually propagated natural progeny still has to be proven. To date, seedlings could not be located in the wild, which does not exclude the possibility of their presence. Further field work is necessary on this subject. It seems likely that in this area the propagation of *A. philoxeroides* is mostly vegetative. Stems thrown on dry surface or in water root readily and form new plants within a few days. In the water the unrooted cuttings keep growing without interruption, though somewhat slower than a well-established plant. On dry ground the roots have to be formed before the plant continues its growth.

Stem tips of less than 0.5 cm to stems of 2 m rooted readily in both media, water and soil. The presence of leaves on the stem was not necessary; cuttings with removed leaves rooted within the same time span as those with their leaves left on. In both media, the callus forms on the severed part of the petiole but root formation does not occur and the leaf disintegrates in a few days. Nodes root in both soil and water; roots appear in approximately 48 hr; stems and leaves emerge from the buds within 72 hr. Internodes do not root in soil but in water. There is root formation in approximately 72 hr, but only on 5% of the cuttings.

SURVIVAL IN TOTAL DARKNESS

Field observation of thick mats of alligator weed led to the assumption that underwater stems covered by a heavy layer of vegetation may stay dormant for a considerable length of time without light.

Strong cuttings of pencil thickness, with two nodes each, were taken and placed in beakers filled with tap water. The beakers were enveloped in a double layer of black polyethylene and sealed with a tape. The beakers were opened after a time lapse of 10, 20, 50, 150 and 365 days at room temperature. The results are shown in Table 11.4.

Alligator weed may survive and increase after being kept 365 days in total darkness.

DISPERSAL OF SEED AND VEGETATIVE PARTS

No proof of dispersal by seed has been found in Louisiana. According to the results of the experiments made with parts of plants, it has been proven that even very small particles of the nodes develop into full-size plants in a very short length of time, expecially during the months with high water and soil temperatures. *A. philoxeroides* is a hydrophilous, ornithophilous, and zoophilous taxon, taking into consideration that even very small particles of the plant have the capability to root wherever they are dropped on water or land.

Floods can carry plants over levees; after the water recedes, these plants become terrestrial thus establishing a new habitat. Two hurricanes dumped sea water which became trapped in the normally faintly brackish marshes (10% of sea-strength salinity) of Cameron Parish. These hurricanes were accompanied by an unusual spring drought, and the evaporated water was not replaced and the marshes came close to having the character of a salt flat. Alligator weed was found growing there in profusion as a halophyte.

The plants were found just before the beginning of the wet season and it seemed important to find out if these plants would survive the sudden change the fresh water would necessarily bring about; sudden disappearance of salinity may lethally damage the tissues. Collected plants were placed in beakers containing water with a salinity of 50% sea strength. The salt water was replaced by flushing with fresh water, within 7 days; the plants did not show ill effects and their habitu gradually changed back to normal.

TABLE 11.4

Survival of Alligator Weed in Total Darkness

Exposure to Darkness (days)	Observation	Date	Followup
10	Roots formed; buds beginning to grow.	6-28-63	Vigorous growth.
20	Good root formation. Uppermost buds grew to lengths of 8 and 15 cm. No chlorophyll apparent.	7-5-63	Chlorophyll apparent. Vigorous growth.
50	Nodes and internodes without chlorophyll. Nodes destroyed. Buds viable.	7-30-64	Healthy plants developed from buds.
100	Good root formation. Buds developed into new shoots 10, 12, and 15 cm long. No chlorophyll apparent.	10-1-64	Chlorophyll apparent. Vigorous growth.
150	Internodes completely destroyed. Only small disklike nodal sections remain with rootlets and healthy buds.	12-1-63	Healthy plants. Growing well.
365	One cutting totally destroyed. One cutting with good root system at lower node, and with a 34-cm-long new shoot. All parts without chlorophyll.	6-20-64	Chlorophyll apparent. Plants growing well.

Chapter 12

FIELD RESEARCH ON CONTROL METHODS*

Under sponsorship of the U.S. Army Corps of Engineers, a multifaceted research program is being conducted on major aquatic weed species in the United States. Alligator weed (*Alternanthera philoxeroides* [Mart.] Griseb.) was an early target of this research due to the difficulty of controlling the weed chemically. Research in Louisiana, where alligator weed is a problem in rice-growing areas and in canals and lakes, resulted in data on the use of four phenoxy herbicides. Included with this is information on the release of the alligator weed flea beetle (*Agasicles hygrophila* Selman & Vogt) in the state, and subsequent impact on alligator weed at several sites, 1959–1972.

INTRODUCTION

Virtually every type of water transportation in the nation's larger inland waterways is affected by dense growths of aquatic plants. Included are small pleasure craft, commercial fishing fleets, petroleum industry vessels, and modern barge tows which move hundreds of important commodities. These plants also increase the chance of local flooding by impeding natural runoff and are detrimental to fish and wildlife in bayous, swamps, and marsh areas adjacent to navigable waterways. They affect agriculture by increasing water loss through evapotranspiration and impeding water management in irrigation canals, thereby increasing production costs.

CHEMICAL CONTROL

Preliminary studies on the chemical control of alligator weed and water hyacinth in Louisiana were made in 1950 under a research contract with the New Orleans District of the U.S. Army Corps of Engineers and Tulane University. Since that time, phenoxy herbicides have been widely used for control of these two aquatic weeds. However, herbicide application for control of alligator weed has been quite limited because relatively high herbicide levels are required, and the cost of retreatment that is usually necessary after a short period of time is substantial. To improve the control method, research studies were initiated

with the University of Southwestern Louisiana in 1959.[15]

Field trials for the summer seasons of 1964, 1965, and 1966 were set out in Louisiana. Randomly selected plots 600 m^2 in surface area were sprayed with phenoxy herbicides using conventional ground equipment. The herbicides were applied in a water solution at a rate of 1.87 kl/ha and 14 kg/cm.[2] Eleven large-scale plots, 0.2 to 0.4 ha in size, were sprayed by Corps of Engineers crews, using the same equipment and 2,4-dichlorophenoxy acid (2,4-D) herbicides used for the control of water hyacinths. These plots were established in the canals of the Bonnet Carre Spillway near New Orleans (see Figure 12.1).

Field data for the summer seasons of 1964, 1965, and 1966 are presented in Tables 12.1, 12.2, and 12.3. The herbicides 2-(2,4,5-trichlorophenoxy)propionic acid (Silvex®), the propylene glycol butyl ether esters (PGBEE), and the dimethyl amine (DMA) formulations of 2,4-D successfully controlled alligator weed. Differences between herbicides were not significant. Rates and dates were highly significant, i.e., applications below 8 kg/ha did not produce satisfactory control, and late summer applications were more effective than early summer applications. Large-scale field trials showed regrowth along the canal banks during the second year, and most canals were covered over by the end of the third year.

BIOLOGICAL CONTROL

Preliminary studies on the successful control of alligator weed with the alligator weed flea beetle have been reported in Florida[14] (see Figures 12.2 and 12.3). There was some opposition to the introduction of the alligator weed flea beetle into Louisiana for biological control of this weed because some economic uses existed, such as cover and food in crayfish ponds and grazing for cattle during the winter season. These uses, however, did not greatly conflict with the use of the alligator weed flea beetle for control of the plant in canals and bayous.

The cooperative investigation of the biological

From Gangstad, E. O., Foret, J. A., and Spencer, N. R., *Hyacinth Control J.,* 13, 30, 1975. With permission.

FIGURE 12.1.   Site of University of Southern Louisiana research plots at the Bonnet Carre spillway near New Orleans, Louisiana, 1964 to 1966. (Photograph courtesy of U.S. Army Corps of Engineers.)

control of this weed was initiated between the U.S. Army Corps of Engineers and the Agricultural Research Service (ARS) of the U.S. Department of Agriculture (USDA) in 1959. Exploration and investigation for insect enemies of the plant in South America were done by personnel of the Systematic Entomology Laboratory of the ARS.[16] ARS personnel conducted host specificity studies at the USDA research station in Argentina and at the USDA-ARS Biological Control Laboratory in California. Results of host specificity studies in Argentina indicated that an undescribed flea beetle, later named the alligator weed flea beetle, was suitable for biological control of alligator weed.[5-8]

In December 1970, an official permit was granted by Louisiana authorities for the release of the insect in the state. The first release was made in February 1971, in the Cross Bayou Canal, about

8 km west of the New Orleans International Airport. Since that time, 15 releases have been made in Louisiana, with control of alligator weed apparent in most areas of the state (see Figures 12.4 and 12.5).

Release sites of the alligator weed flea beetle in Louisiana are summarized in Table 12.4. Most of these sites have active populations of the alligator weed flea beetle, and some measure of control has been obtained. The beetles are found to be most active during the spring and fall with a slowing of activity during the hot summer months. They have overwintered in the southern part of the state, and indications are that a stable population has been established. There is little doubt that the alligator weed flea beetle will continue to suppress excessive growth of alligator weed growing in aquatic situations but terrestrial growth is much less affected.

TABLE 12.1

Percent Control for Herbicide Treatments at Different Rates and Months of Application for Control of Alligator Weed for April, May, and June 1964

Herbicide treatment[a]	Rate (kg/ha)[b]	Month of application	% Control[c]
Silvex®	8.4	April	80[d]
	5.0	April	55
	2.7	April	15
2,4-D (PGBEE)	8.4	April	80[d]
	5.0	April	70
	2.7	April	25
2,4-D (PGBEE)	8.4	May	85[d]
	5.0	May	45
	2.7	May	30
2,4-D (PGBEE)	8.4	June	73[d]
	5.0	June	40
	2.7	June	30

Summary of Treatment Means

Herbicides			
Silvex			50[e]
2,4-D (PGBEE)			52
Months		April	54[e]
		May	53
		June	48
Rates			
	8.4		79[d]
	5.0		52
	2.7		30

[a]Applied in a water solution at 1.87 kl/ha and 14 kg/cm^2.
[b]Rate in kilograms of active ingredients per hectare.
[c]Percent control estimated from three replicates by visual inspection of below water mat, 6 weeks after treatment (Weed Science Society of America rating 1–10).
[d]Significant difference, P = 0.02.
[e]Nonsignificant difference.

INTEGRATED CONTROL

As early as June 1968, Zurburg[15] suggested that the improved control of alligator weed in Florida at the Ortega River release site might be related to the combined activity of the alligator weed flea beetle feeding on the above portion of the mat and the effect of 2,4-D (incidentally applied for control of water hyacinth) on the underwater portion of the mat. It was not until after permission was granted for the introduction of the alligator weed flea beetle by authorities in Louisiana that this approach to the problem could be further studied in Louisiana.

Field observations have been made since 1970 to determine the effectiveness of the alligator

TABLE 12.2

Percent Control of Herbicide Treatments for Three Levels of Application for Control of Alligator Weed for May and July 1965

Herbicide treatment[a]	Rate (kg/ha)[b]	% Control[c] May	% Control[c] July
Silvex®	8.4	70	100
2,4-D (PGBEE)	8.4	50	100
Silvex	5.0	35	60
2,4-D (PGBEE)	5.0	50	100
Silvex	2.7	35	60
2,4-D (PGBEE)	2.7	20	60

Summary of Treatment Means

Herbicides			
Silvex		47	73[d]
2,4-D		40	88
Rates			
	8.4	60[a]	100[e]
	5.0	43	80
	2.7	26	60
Months			
May vs. July		43	80[e]

[a]Applied in water solution at 1.87 kl/ha and 14 kg/cm^2.
[b]Rate in kilograms of active ingredient per hectare.
[c]Percent control estimated from three replicates by visual inspection of the below water mat, 6 weeks after treatment (Weed Science Society of America rating 1–10).
[d]Nonsignificant difference.
[e]Significant, P = 0.05.

From Gangstad, E. O., et al., *Hyacinth Control J.,* 13, 31, 1975. With permission.

weed flea beetle to control alligator weed. During the initial period of dispersion, populations increased slowly; consequently, feeding damage was minimal. However, by the fall of 1971 significant populations and feeding were apparent. As a result of the mild 1971 to 1972 winter, insects overwintered and large populations were found in most of the state in the spring of 1972.

In several widely separated locations in Louisiana during 1971 and 1972, attack by the alligator weed flea beetle on alligator weed was followed by increased competition from water hyacinths and eventual replacement of the alligator weed by water hyacinth. In these cases, herbicide treatment of water hyacinth resulted in the elimination of both plants.

Approximately 50 adult beetles were released at this site near Lake Pontchartrain on February

TABLE 12.3

Percent Control for Herbicide Treatments of Herbicide Combinations for Control of Alligator Weed for May, July, and September 1966

Herbicide treatment[a]	Rate (kg/ha)[b]	Month of application	% Control[c]
2,4-D (PGBEE) plus Silvex®	4.5	May	25
	2.3	July	40
		September	100[d]
2,4-D (DMA) plus Silvex	4.5	May	30
	2.3	July	25
		September	100[d]
2,4-D (PGBEE)	6.8	May	25
		July	20
		September	100[d]
2,4-D (DMA)	6.8	May	35
		July	30
		September	100[d]

Summary of Treatment Means

Combinations			
2,4-D (PGBEE) plus Silvex			55[e]
2,4-D (DMA) plus Silvex			55
2,4-D (PGBEE)			46
2,4-D (DMA)			36
Months		May	26
		July	29
		September	90[d]

[a] Applied in water at 1.87 kl/ha and 14 kg/cm².

[b] Rate in kilograms of active ingredient per hectare.

[c] Percent control estimated from three replicates by visual inspection of the below water mat, 6 weeks after treatment (Weed Science Society of America rating 1–10).

[d] Significant difference, P = 0.05.

[e] Nonsignificant difference.

From Gangstad, E. O., et al., *Hyacinth Control J.,* 13, 32, 1975. With permission.)

19, 1971. This location was selected as a release site for the alligator weed flea beetle because of the lush growth of alligator weed. The beetles also had access to large quantities of alligator weed growing in the marsh area. Because of the slow increase in population an additional 2,000 beetles were released on April 15, 1971. This release resulted in a rapid increase in the population and in feeding damage (Table 12.5).

In July 1971, water hyacinth intruded into the area and rapidly multiplied throughout the rest of 1971 and into the spring of 1972, to the point that by June 1972 the canal was completely blocked. The canal was sprayed on June 26 and 27, 1972, by crews of the Louisiana Wildlife and Fisheries Commission, using 4.5 kg/ha of the DMA formulation of 2,4-D in a 0.5% spray solution. Spraying of water hyacinth was done without any appreciable damage to the alligator weed. By October 1972, only a small amount of alligator weed remained in the area, and by the beginning of December 1972, alligator weed was no longer present. Continuing surveys in the spring of 1973 indicated that almost no alligator weed was present in the canal and along its banks.

In an evaluation of the causes leading to the severe reduction of alligator weed on Cross Bayou, several factors were considered. The winter of 1971 to 1972 included 4 days below 0°C in January and 3 days in February. This cold weather killed a portion of the alligator weed above the water and delayed its spring growth. In the spring of 1972 the alligator weed flea beetle continued the stress which the weed has been under due to the cold weather earlier in the year. The alligator weed in Cross Bayou was less competitive with

FIGURE 12.2.   Ortega River, Florida. Alligator weed flea beetle release site, looking west, May 26, 1966. This was the first successful release site in the United States. (From Leiger, C. F., *Hyacinth Control J.,* 6, 31, 1967. With permission.)

water hyacinth, thus a plant replacement occurred making the use of 2,4-D necessary. Drift from this spray may have contributed to the disappearance of the alligator weed in the summer of 1972. Alligator weed has not reestablished on the canal due to the continued pressure of biotic and abiotic stresses.

Attempts to control rooted and floating mats of alligator weed emerging from the banks of rice irrigation canals began in April 1968. Combination of rapidly degradable phenoxy herbicides and water management were first successfully used in the 1969 to 1970 season. These treatments, while not completely satisfactory in degree of control obtained, gave adequate control to minimize interference with irrigation water movement. In late October 1970, large-scale spraying of irrigation canals was accomplished by applying 4.5 kg/ha of the DMA formulation of 2,4-D in 30.3 l of water by commercial crop spraying aircraft. This applica-

tion was timed to coincide with lowering water level in the irrigation system and earlier experience with the control of floating mats of alligator weed by fall application of phenoxy herbicides. Results obtained earlier in this study in the control of rooted alligator weed in the irrigation canals demonstrated the advisability of an early spring application (late March) about 1 week prior to filling the canals with muddy water. In July 1971, very little alligator weed was in evidence in the treated canals (approximately 16 ha treated). At this time, the alligator weed flea beetle was introduced on alligator weed along the canal bank and was noted to be feeding on the surviving alligator weed. Beetle damage was moderate during this first summer and a population increase was observed. In the spring of 1972, the alligator weed flea beetle damage to alligator weed was very much in evidence and the remaining stand of alligator weed was practically eliminated.

FIGURE 12.3.   Adult alligator weed flea beetle eating on stems of alligator weed. (From Leiger, C. F., *Hyacinth Control J.,* 6, 31, 1967. With permission.)

FIGURE 12.4.   Release site near Lake Pontchartrain; entomologist Neal R. Spencer, ARS, and aquatic weed expert William E. Thompson, U.S. Army Corps of Engineers, check for egg clusters – an index of *Agasicles* population. (Photograph courtesy of USDA, ARS.)

FIGURE 12.5. Federal and Louisiana state officials making an inspection tour on a waterway that until introduction of *Agasicles* was clogged with alligator weed. The weed will continue to grow at shallow stream banks – a boon to wildlife and livestock. (Photograph courtesy of USDA.)

TABLE 12.4

Alligator Weed Flea Beetle Release Sites in Louisiana

1971 date of release	Number released	Origin of alligator weed flea beetle	Release site
19 February	50	Ortega River, Fla.	Cross Bayou Canal 1
3 March	750	Ortega River, Fla.	Jasmine Bayou 1
3 March	750	Ortega River, Fla.	Cross Bayou Canal 2
19 March	150	Lake Arthur, La.	Jasmine Bayou 2
15 April	2000	Lake Charles, La.	Cross Bayou Canal 3
16 April	1000	Lake Charles, La.	Jasmine Bayou 3
16 April	300	Lake Charles, La.	Bayou Rapids 1
17 April	300	Lake Charles, La.	Brushy Bayou
17 April	300	Lake Charles, La.	Round-a-way Bayou
26 April	500	Ortega River, Fla.	Warren Canal
26 May	300	Lake Charles, La.	Bayou Rapides 2
27 May	250	Lake Charles, La.	Lake Bruin
27 May	550	Lake Charles, La.	Logansport
27 May	1500	Lake Charles, La.	Black Bayou Lake
28 May	150	Lake Charles, La.	Chaplin Lake

Information supplied by W. E. Thompson, U.S. Army Engineer District, New Orleans, and L. V. Richardson, Louisiana Wildlife Fisheries Commission, for the estimated number of adult beetles at each site release.

From Gangstad, E. O., et al., *Hyacinth Control J.*, 13, 32, 1975. With permission.

TABLE 12.5

Percent Control Data for Louisiana Locations for Beetle Population and Alligator Weed Mat Width as Related to Herbicide Treatment

Location of alligator weed	Date of observation	Beetle population per square meter (adults)	Mat width (m)	Mat control (%)
Control canal	5/20/71	0	15	
	6/26/72	0	17	
				0
Cross Bayou Canal, Louisiana[a]	5/20/71	13	23	
	6/26/72	25	6	
				+74
Rice Irrigation Canal, Louisiana[b]	7/14/71	11	21	
	7/16/72	18	3	
				+78

[a]Sprayed with 2,4-D (DMA) for water hyacinth control.
[b]Sprayed with 2,4-D (PGBEE) for alligator weed control.

Chapter 13

BIOLOGICAL CONTROL PROGRAM*

Insects have been used successfully as a form of biological control to suppress alligator weed (*Alternanthera philoxeroides* [Mart.] Griseb.) in Florida and other states of the Southeast under the U.S. Army Corps of Engineers aquatic control program in cooperation with the Division of Entomology Research of the United States Department of Agriculture. The alligator weed flea beetle (*Agasicles hygrophila* Selman & Vogt) was the first host-specific insect to be introduced. Other insects, alligator weed thrips (*Amynothrips andersoni* O'Neill) and a stem-boring moth (*Vogtia malloi* Pastrana) are also host-specific and have been introduced for alligator weed control. Infestations of alligator weed are reduced to a negligible population in most situations in the southeastern states where these insect controls have been released.

INTRODUCTION

The initial responsibilities for and interests in aquatic plant control of the Corps of Engineers arose from the widespread and profuse growths of alligator weed and water hyacinths (*Eichhornia crassipes* [Mart.] Solms) that limited navigation in rivers and harbors of the southern states. When these aquatic plant infestations constitute a serious economic threat to navigation, flood control, drainage, agriculture, water quality, and related purposes, control projects are authorized within budgetry limitations set by the Congress of the United States in 1965. Mechanical methods were used during the first phase of this program,[12] chemical methods in the second phase,[4,13] and biological control as a third phase.[5-8,14]

EXPLORATION FOR NATURAL ENEMIES

A research agreement was initiated by the Corps of Engineers in 1959 with the United States Department of Agriculture, Division of Entomology Research, to conduct explorations in South America for natural enemies of alligator weed. George B. Vogt, research entomologist with the Systematic Entomology Laboratory, Washing-

ton, D.C., made the original survey and observed a flea beetle on alligator weed in its natural habitat (Figure 13.1). A research laboratory was set up in Argentina for further research. The principal and immediate objective of the research program was to determine whether or not the alligator weed flea beetle could complete its life cycle on any plant except its normal host. All evidence from these studies indicated that the alligator weed flea beetle is an obligatory monophagous insect, and is the principal suppressant of alligator weed in its native habitat.[16]

HOST SPECIFICITY

Laboratory feeding studies were conducted on *Polygonum, Fagopyrum, Rheum, Chenopodium, Atriplex, Amaranthus,* and *Alternanthera.* These feeding tests also included *Oryza, Nasturtium,* and *Nymphaea. Atriplex hastata* L. was the only species other than alligator weed which was fed upon by the beetle and this species did not permit completion of the life cycle. In the feeding experiments with *A. hastata* both larval and adult flea beetles fed on the test plant. In two of the experiments, larvae fed on the leaves of *A. hastata* and development at first appeared to be normal when compared with the controls. However, by the third day of the tests, the larvae became restless and exhibited migratory tendencies. Some of the larvae died. At the end of the eighth day all larvae were dead. In the third larval test, feeding was also evident, and three of the five larvae completed their development and pupated in a glass tube. Two of these pupae died but the third pupa was metamorphosed to an abnormal, malformed adult which died within a few hours. In the experiments with adults, feeding was observed in all three experiments but was confined to the stems of the plant. As a result of this feeding the adults lived an average of 21.7 days but showed abnormal behavior and died without producing eggs. In fact, dissection of the females subsequent to death demonstrated no ovarian development. Furthermore, *A. hastata* does not have a hollow stem, required in nature as a site for pupation. Observations of the plant under growing condi-

From Gangstad, E. O., *Aquatic Plant Management,* 14, 50, 1976. With permission.

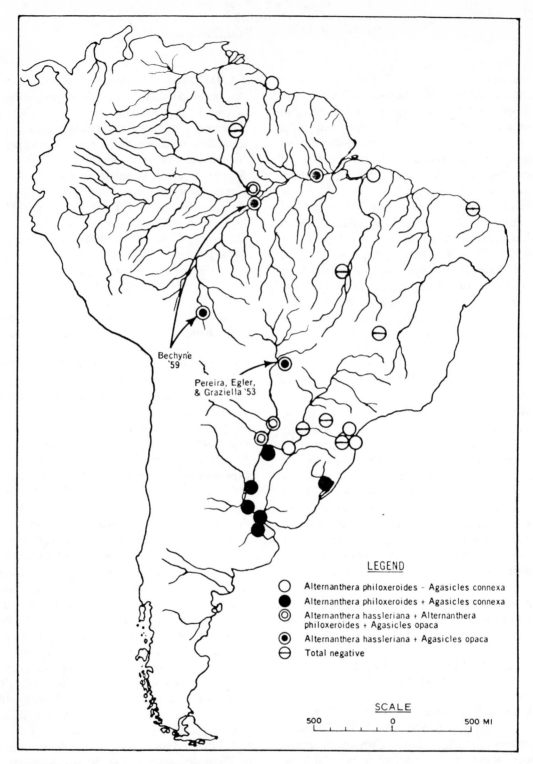

FIGURE 13.1.   Coverage of alligator weed survey in South America, 1953 to 1959, for insect enemies.

LEGEND

○ Alternanthera philoxeroides − Agasicles connexa
● Alternanthera philoxeroides + Agasicles connexa
◎ Alternanthera hassleriana + Alternanthera philoxeroides + Agasicles opaca
⦿ Alternanthera hassleriana + Agasicles opaca
⊖ Total negative

SCALE

500   0   500 MI

tions have failed to demonstrate feeding by either larvae or adults of the flea beetle.[5-8] Plants which were resistant to attack of the alligator weed flea beetle are given in Table 13.1.

INSECT INTRODUCTION

The decision was made that this flea beetle w the most promising biocontrol for alligator wee and the insect was brought, under quarantine, to the research laboratory in Albany, California, for further study in 1963. The first release was made in South Carolina in 1964 at the Savannah National Wildlife Refuge under conditions which appeared to be similar to those in South America. These early results were not very impressive. The first successful biological control was observed in the Ortega River, near Jacksonville, Florida, in

TABLE 13.1

Plants Which are Resistant to Alligator Weed Flea Beetle Attack

I. Polygonaceae (Buckwheat family)
   A. *Polygonum*
      1. *P. aviculare* L. – Common or yard knotweed; annual.
      2. *P. hydropiperoides* Michx. – Mild water pepper; perennial.
      3. *P. punctatum* Elliot. – Water smart-weed; perennial.
      4. *P. densiflorum* Mesin. – (densely flowered); perennial.
   B. *Fagopyrum*
      1. *F. sagittatum* Gilib. – Common buckwheat.
   C. *Rheum*
      1. *R. rhaponticum* L. – Rhubarb; perennial.
II. Chenopodiaceae (Goosefoot family)
   A. *Chenopodium*
      1. *C. macrospermum* Hook. F. var. *farinosum* (Wats.) – Annual.
      2. *C. ambrosioides* L. – Mexican tea; Short-lived perennial.
   B. *Atriplex*
      1. *A. hastata* L. – Saltbush; annual.
      2. *A. hortensis* L. – Garden orache; annual.
      3. *A. semibaccata* R. Br. – Australian saltbrush; perennial.
III. Amaranthaceae (Pigweed family)
   A. *Amaranthus*
      1. *A. defluxus* L. – Low amaranth; annual.
      2. *A. standleyanus* Parodi – annual.
      3. *A. lividus* L. var. *ascendens* (Lois.) Thell. – Annual.
   B. *Alternanthera*
      1. *A. bettzichiana* (Reg.) Standl. – Ornamental perennial.
      2. *A. pungens* H.B.K. – Yerva Del Pollo; perennial.
      3. *A. repens* (L.) Kuntze. – Perennial.
IV. Miscellaneous
      1. *Oryza sativa* L. – Rice; annual (Graminaea).

From Gangstad, E. O., *J. Aquatic Plant Management*, , 51, 1976. With permission.

1965. Most of the alligator weed flea beetles distributed in the United States have come from this area.[8] Conditions in Florida were apparently more favorable than elsewhere. Successful control, however, may have been due in part to the effects of (2,4-dichlorophenoxy)acetic acid (2,4-D) on the alligator weed mat from incidental treatment of water hyacinth, shortly after release of the beetle.[17]

DISTRIBUTION IN SOUTHEASTERN UNITED STATES

During May 1965, two trips were made into the southeastern United States to distribute the alligator weed flea beetle and to initiate evaluation studies. The first trip was made by Dr. W. H. Anderson of the U.S. Department of Agriculture on May 7 to 12. His primary objective was to release beetles at selected localities in Georgia, South Carolina, and North Carolina. On May 9, with Charles Zeiger and James McGeehee of the Corps of Engineers, Jacksonville, he visited the original release site on the Ortega River. The alligator weed infestation at this site showed no evidence of recovery; there were only scattered plants growing along the banks and a few small floating islands that had drifted in from the river. At another site on Black Creek south of Jacksonville, where there was a considerable amount of alligator weed in small as well as extensive patches along the banks, all plants were found to be under heavy attack. Some of the patches were entirely brown with the stems prone and badly chewed.

Additional releases were made at the Jim Woodruff Reservoir in Georgia and Florida, at three sites near Mobile, Alabama, at Gulfport and Yazoo City, Mississippi, at two sites on the Dam B reservoir near Jasper, Texas, and at three sites in the J. D. Murphree Wildlife Area near Port Arthur, Texas. The general distribution is summarized in Table 13.2.

VEGETATIVE CONTROL

Throughout its life stages, the flea beetle attacks the alligator weed in different ways. The adults feed on surface leaves; females lay their eggs – 1000 or more – on the undersides of leaves; young larvae then feed on the undersurface of the leaf and, as mature larvae, chew their way into the

TABLE 13.2

Distribution of Alligator Weed Flea Beetles in North and South Carolina, Georgia, Florida, Alabama, Mississippi, Texas, and Tennessee

Name and address of cooperator	Location of site

North Carolina

Mr. O. H. Johnson
  U.S. Army Engineer District,
  Wilmington, N.C.
Mr. Jessie Sessions
  Office of the State Entomologist
  State Department of Agriculture
  Wilmington, N.C.

Chadbourn, released flea beetle
  in a farm drainage ditch about 1000
  ft from the intersection of country
  roads 1560 and 1562
At marker separating Brunswick
  and New Hanover Counties
On property owned by Time Corp.
Lake Waccamaw
Vicinity of Wilmington

South Carolina

Mr. Jack J. Lesemann
  Chief, Engineering Division
  U.S. Army Engineer District
  Charleston, S.C.

Lake Marion, Santee; southwestern
  shore between towns of
  Elloree and Lone Star. Released
  at bridge in swamp (Halfway Swamp)
Ashepoo River where it crosses
  the eastern alternate of U.S. 17
On the grounds of the Vegetable
  Breeding Laboratory at Charleston
Goose Creek Reservoir near
  building occupied by reservoir
  personnel
Edisto River, where U.S. Route
  78 crosses river
Black River at Kingstree
Naval facility at Charleston

Georgia

Mr. Angus K. Gholson
  Jim Woodruff Reservoir
  U.S. Army Corps of Engineers
  Chattahoochee, Fla.

Savannah National Wildlife
  Refuge, Pool 3
Mouth of Ebenezer Creek on
  Savannah River
Casey Canal, Savannah
Jim Woodruff Reservoir on
  Flint River arm

Florida

Mr. Charles F. Zeiger
  U.S. Army Engineer District
  Jacksonville, Fla.

Release on the Ontega River
  and approximately 50 other
  locations Florida

Alabama

Mr. W. E. Ruland
  U.S. Army Engineer District
  Mobile, Ala.

Perch Creek, site 1, near church
Perch Creek, site 2, near bridge
Three Mile Creek, near bridge
On canal crossing Halls Mill

TABLE 13.2 (continued)

Distribution of Alligator Weed Flea Beetles in North and South Carolina, Georgia, Florida, Alabama, Mississippi, Texas, and Tennessee

Name and address of cooperator	Location of site
Alabama (continued)	
Mr. George Allen U.S. Army Engineer District Mobile, Ala.	On Black Warrior River at the Dempolis Reservoir Gulf Shores, in the State Park
Mississippi	
Mr. Milton F. Parkman U.S. Army Engineer Division, Vicksburg, Miss.	Yazoo River, Yazoo City White Sand Creek, Prentiss Keyser Bayou, Gulfport
Texas	
Mr. Robert N. Hambric Texas Parks and Wildlife Department Houston, Tex.	Dam B. Reservoir: Site 1, Walnut Ridge Park Site 2, Bridge on Highway from Jasper to Livingston
Mr. Charles D. Stutzenbaker Texas Parks and Wildlife Department Port Arthur, Tex.	J. D. Murphree Wildlife Area; Site 1, Taylor Bayou Air-boat Trail
Mr. Clifford J. Novosad U.S. Army Engineer District Galveston, Tex.	Site 2, Outside ditch, Compartment 1, at hydro-flow gate; Site 3, Mouth of Deering Slough
Tennessee	
Dr. Gordon E. Smith Tennessee Valley Authority Mussel Shoals, Ala.	At Mussel Shoals and other locations in Tennessee

From Gangstad, E. O., *J. Aquatic Plant Management,* 14, 52, 1976. With permission.

stems. Larvae develop into adults within the stems, eat their way out, and return to the leaves to start the cycle again. Damage by the beetle and larvae either kills the alligator weed outright, or weakens it, making it vulnerable to disease, competition from other aquatic plants, and wind and wave action.

The program for biocontrol of alligator weed with the flea beetle has been generally satisfactory within the limits expected. The estimated acreage of infestation and acreage of chemical treatment of the Corps program is summarized in Table 13.3. The acreage of infestation was reduced from 97,186 acres in 1963 to 78,030 acres in 1973. The acreage for treatment with herbicides was reduced from 21,805 acres in 1963 to 5,594 acres in 1973. The current results indicate that alligator weed has been reduced to a negligible population in most infestations in the southeastern states where insect controls have been released.

TABLE 13.3

Acreage of Infestation and Chemical Treatment of Alligator Weed for 1963 to 1973[a]

Year	Area of infestation		Area of treatment	
	1963 (acres)	1973 (acres)	1963 (acres)	1973 (acres)
South Atlantic division				
Jacksonville, Fla.	2,597	Minor	50	None
Savannah, Ga.	1,838	Minor	50	None
Wilmington, N.C.	428	3,220	100	235
Charleston, S.C.	30,430	29,710	750	750
Mobile, Ala.	4,813	225	50	109
Lower Mississippi Valley division				
New Orleans, La.	55,880	36,275	19,605	4,000
Vicksburg, Miss.	None	200	None	200
Southwestern division				
Galveston, Tex.	1,200	8,400	1,200	300
Total acreage	97,186	78,030	21,805	5,594

[a]Estimate of acreage by field crews.

From Gangstad, E. O., *J. Aquatic Plant Management,* 14, 53, 1976. With permission.

Chapter 14

INTEGRATED CONTROL PROGRAM*

Results of laboratory and field experiments have shown that the use of introduced insects alone frequently does not result in a satisfactory reduction of floating alligator weed mats, and application of herbicides is necessary. Although the use of insects alone may result in an acceptable level of control over an extended period of time, if the reduction of alligator weed is the immediate objective, then an integrated approach should be used. The following is a detailed outline for a 3-year period to illustrate the use of an integrated approach for alligator weed management along the southeastern coast of the United States.

INTRODUCTION

Alligator weed infestations have thrived and increased in the southeastern states for nearly a century. The plants grow profusely in cultivated fields, drainage ditches, canals, rivers, lakes, and reservoirs. The first recorded problem was in 1897 when this plant completely filled a creek near Mobile, Alabama.[10]

Alligator weed varies in different locations, growing as a terrestrial, rooted-immersed, or completely free-floating plant. When free floating, it forms a dense interwoven mat of stems.

Floating mats create the greatest problems for navigation and drainage[10] and are difficult to control with herbicides. Multiple applications of 2-(2,4,5-trichlorophenoxy)propionic acid (Silvex®) and (2,4-dichlorophenoxy)acetic acid (2,4-D) will reduce alligator weed mats somewhat;[11] however, the economics of repeated applications limit the use of this control method.

The flea beetle (*Agasicles hygrophila* Selman and Vogt), thrips (*Amynothrips andersoni* O'Neill), and moth (*Vogtia malloi* Pastrana) have been introduced as biological control agents for alligator weed and are well established in many areas in the United States.[1] The flea beetle, which was introduced first, appears to be the most effective insect.

The effectiveness and spread of the flea beetle has been monitored since its release in 1964. In certain areas, peak flea beetle activity occurs twice annually, and in these areas the flea beetles have demonstrated an ability to severely damage surface vegetation of floating alligator weed.[17] In Northern Florida, the two peaks of flea beetle activity, aided by plant competition from water hyacinth (*Eichhornia crassipes* [Mart.] Solms) and other aquatic plants, has reduced the size of alligator weed infestations. In Georgia and South Carolina few flea beetles overwinter. Evidence of flea beetle activity is not noticeable until late summer or early fall. This corresponds with the period prior to frost when indigenous species of insects sometimes produce severe damage to surface alligator weed.

Alligator weed growth rates are greatest in early spring. New shoots continue to elongate until the plants fall over and become incorporated into the subsurface mat. The timing of this event varies with location, but usually occurs in midsummer. Effectiveness of alligator weed control will be determined by the degree of flea beetle feeding activity on the surface vegetation. A constant pressure applied by insect feeding on the plant prevents subsurface build-up of floating mats.

Results from experiments on small plots in the field[18] indicated that combined chemical-biological control is more effective than control obtained with either agent alone. The success of the integrated control program appears to be related to the prevention of plant replacement in subsurface mats.

PROGRAM FOR ALLIGATOR WEED MANAGEMENT

The degree of control desired determines the most suitable method for alligator weed management. The suggested program (see Figure 14.1) is based on research conducted in greenhouse and field experiments in Florida, Georgia, and South Carolina. In coastal rivers in the southeastern states floating alligator weed mats may not be controlled by the program in some cases because there is a mechanical movement of mats from canals or impoundments into the river system. If evidence is found of alligator weed mat discharge, it is advisable to initiate a control program in the

rom Gangstad, E. O., Blackburn, R. D., and Spencer, N. R., "Integration of Biological and Chemical Control of Alliga-weed," Cooperative study of the U.S. Army Corps of Engineers and the U.S. Department of Agriculture, Fort .uderdale, Florida, unpublished report, 1973.

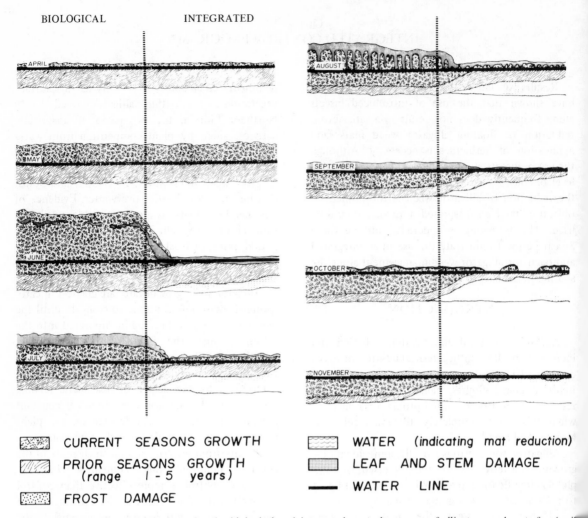

BIOLOGICAL   INTEGRATED

APRIL
MAY
JUNE
JULY

AUGUST
SEPTEMBER
OCTOBER
NOVEMBER

▨▨ CURRENT SEASONS GROWTH

▨▨ PRIOR SEASONS GROWTH
    (range 1 - 5 years)

▨▨ FROST DAMAGE

▨▨ WATER (indicating mat reduction)

▨▨ LEAF AND STEM DAMAGE

▬▬ WATER LINE

FIGURE 14.1. Sequence of events for the biological and integrated control program of alligator weed mats for April through November. (From Duren, W. C., Jr., Blackburn, R. D., and Gangstad, E. O., *Hyacinth Control J.*, 13, 27, 1975. With permission.)

areas which provide the source of infesting plant material.

STEPWISE MANAGEMENT
CONTROL PROGRAM

A. Program for the first year
  1. February to March
    a. Survey alligator weed infestations in problem areas.
    b. Use topographic maps where treatment is desired.
    c. Determine location and approximate size of floating mats.
  2. April
    a. Select several mats of alligator weed to be used in monitoring growth rates of sur-

face vegetation. Mats near a boat landing or road bridge are ideal because of easy accessibility.
    b. Tag 25 plants with 0.5-in.^{2} plastic tags. Insert small wire of the type used in aluminum screens with tag attached, through the top of the fourth internode.
    c. Loosely encircle tagged plants with number 14 copper wire and attach to plastic bottle.
    d. Secure bottle by inserting attached wire through a portion of the floating mat. The bottle will be used to locate tags when making observations.
  3. May to June
    a. Observe tagged plants in the floating alligator weed mats that were established

in April as references for plant growth rate. As stem elongation occurs, the tags will be found in a lower position within the surface vegetation.

b. Determine flea beetle and other insect feeding damage on leaves and stems.

4. June to July
a. Initiate herbicide application when tagged internodes begin to fall over onto the floating mat. Application of herbicide at this time will kill most of the surface vegetation produced during the current growth season.

b. Treat floating mats with the dimethylamine formulation of 2,4-D at the rate of 2 lb active ingredient per acre (2.24 kg acid equivalent per hectare). The herbicide should be applied evenly over the plant foliage. Spray volume and pressure used when applying the herbicide will be determined by the type of operational equipment and area.

c. Do not spray areas where flea beetles have congregated in large numbers. The skipped areas and most of the areas missed during herbicidal application will be controlled by insects as they migrate from vegetation that has been treated.

5. August
a. Evaluate the effectiveness of the herbicidal application after 4 to 6 weeks. Regrowth in treated areas may have an average height of 6 in. If insects were noted prior to spraying, then feeding damage should be evident on regrowth.

b. If no insect damage is noted on regrowth, then collect and release introduced insects on regrowth. The number released can vary, but should average 100 for each acre treated. Flea beetles and alligator weed moths should be released because of their mobility and ability to severely damage regrowth of treated alligator weed mats. Note: Although introduced insects are difficult to find early in the spring season, they are usually abundant near late summer, depending on previous winter temperatures. Insects usually can be collected in previously infested areas.

6. December
a. Observe and evaluate the effectiveness of the management program in the reference areas. Frost should have already occurred

in the area and plastic tags placed in the floating mat in April can be easily located. Internodes with tags attached should be dead or in a poor condition.

b. Record approximate reduction in size of floating mats and the general effectiveness of the program. If tags are found with dead internodes attached this is proof that most of the season's surface vegetation has been killed, thus preventing mat replenishment. The floating mats should be in poor condition, and the alligator weed infestations should have been reduced a minimum of 50%.

B. Program for the second year
7. March
a. Survey alligator weed infestations and determine the percent reduction.

b. If the reduction is satisfactory, it is suggested that Step 5b be repeated. A small increase in mat size may occur during the spring and summer months.

c. Should additional reduction in alligator weed mats be desired, Steps 4b, 4c, 5a, and 5b should be repeated. This should result in a minimum of 80% reduction in alligator weed by the end of the second year. Tagging of alligator weed plants the second year will be optional. The approximate time of application should have been established in Year 1. However, if tags were found attached to thriving internodes in December (Step 6a), then initiate the herbicidal application 2 to 3 weeks earlier than the previous year.

8. September to November
a. Examine plant material and determine the population and effectiveness of the insects.

b. Record approximate alligator weed infestation.

C. Program for the third year
9. March
a. Repeat Steps 7a and 7b.

b. If Step 8 was used during the second year and the alligator weed mats have increased beyond acceptable control, then repeat Step 4b.

DESCRIPTION OF EVENTS

To illustrate the biological and integrated methods of control, the sequence of events is pictured in Figure 14.1. Left of center the potential effects of insects when used as biological control agents are indicated. Right of center are revealed the effects of an integrated approach where herbicide application is followed by insect feeding. The events are summarized as follows:

- April: Plant height averages 12 in. and is well established. Growth rate of alligator weed is increasing rapidly with little or no insect damage to the plant.
- May: Plant height has increased to 18 in. Plant damage by insects is light. The weight of surface vegetation is increased but is not sufficient to press existing growth into the mat.
- June (biological): Insect populations have increased and insect feeding is evident on a few mats along the river. Feeding damage will be more severe near the mat perimeter. Plant height averages 18 in. Additional surface weight will have pressed early growth below the water, thus adding to subsurface mat.
- June (integrated): Areas are monitored and surface vegetation sprayed with herbicide prior to plant fallover. This eliminates the surface growth that would normally be added to the subsurface mat.
- July (biological): Growth rates decrease, but stems continue to elongate and height of plants may average 20 in. Close examination reveals stems of current season's growth well below the water line. Insect damage is found on most leaves, but damage usually ranges from light to moderate and does not affect the growth substantially.
- July (integrated): Areas which have been treated 1 month previously will have regrowth averaging 2 to 4 in., and if insects are present they will have damaged the leaves and stems. The leaves of regrowth will be small and therefore more susceptible to insect damage.
- August (biological): Height of plants will have decreased because of added surface weight and plant fallover. Attack by domestic insects along the introduced flea beetles and moths will have damaged most of the leaves and stems.

Damage by insects will be more predominant near the mat perimeter.

- August (integrated): The effect of insect feeding on regrowth from previously treated mats will be severe. Plant height will vary because some alligator weed stems will have shriveled and died.
- September (biological): Fall peak of insect activity will have reduced surface vegetation. Internodes will be void of leaves and dead. Plants of current season's growth which have been pressed below the water line will not be affected by insect damage.
- September (integrated): Regrowth will be reduced back to the water line by insects. The internodes below the water will be yellow and brittle because they will represent growth produced the previous season. Floating mats will be thin due to the combined pressures of chemical application and insect feeding damage.
- October (biological): Regrowth from attacked plants will be sparse. Insect populations will have decreased because the adults have left in search of alligator weed with less damage. Insects that have remained will prevent mat replenishment.
- October (integrated): Regrowth will be sparse. Internodes near the surface will be limp and void of chlorophyll. Deterioration of the mat will have advanced to the point that some mats will break free and float downstream. Mat fragments caught in the bend of the river will be rolled and inverted because of water pressure.
- November (biological): Frost will have killed surface vegetation. New leaves will be produced by stems pressed into the mat during the growing season. However, chilling temperatures will prevent any substantial growth. Insects may be found near the shore or under the cover of dried vegetation or floating on the surface. When first examined the insects will appear dead, but when placed in the sun they will begin to revive.
- November (integrated): Mats will be reduced by the combined pressures of herbicide and insect feeding damage. Large mats will have holes where portions of the mat have dropped out because of deterioration of internodes. Most of the alligator weed infestation will be marginal.

Chapter 15

IMPACT ASSESSMENT AND FEEDING
DAMAGE OF INSECTS ON ALLIGATOR WEED*

Obnoxious aquatic plants such as alligator weed may completely block a channel to navigation, impede water flow as much as 80%, and have an adverse impact on project objectives, purposes, and operations. The use of canonical correlations has enabled us to more accurately assess the interrelationships which exist between insects introduced for control of alligator weed. The stress of insect feeding on alligator weed increases directly with increased attack, resulting in plants with shorter stems and smaller stem diameters. A second relationship indicates an overall reduction in plant height. As stem density increases, the tendency for *Vogtia malloi* Pastrana to occur decreases and the tendency for *Agasicles hygrophila* Selman & Vogt to occur increases. These interactions tend to be complementary, and the result is a more complete biological control of alligator weed over a wider area of infestation.

INTRODUCTION

Alligator weed grows as a terrestrial plant along canals, streams, and lake shores, or as an immersed aquatic plant rooted near the shore. It forms floating mats of interwoven stems, frequently extending 30 m or more out over the surface of the water. Most aquatic plants are unable to compete with alligator weed in areas where it has been introduced, and they are quickly crowded out. The degree of problem constraint of alligator weed on water resource projects is evaluated by an impact assessment as presented in Table 15.1.

The interactions and relationships which exist between the different types of alligator weed and populations of the phytophagous insects *Vogtia malloi* Pastrana and *Agasicles hygrophila* Selman & Vogt introduced for biological control are known to be significant. To study these associations, a multivariate analysis known as canonical correlations was used.[3] In canonical correlations, a set of independent variables may be compared with a set of dependent variables in such a way as to find the linear combination of variables in each set which, when correlated, is maximum. The resultant

variable is known as a canonical variate. If some linear relationship between the sets of variables still remains unaccounted for by the first set of canonical variates, we can continue the process of finding new linear combinations that would best account for the residual relationships between the two sets. This process can go on until there are no significant linear associations left. In this program of analysis each canonical variate is orthogonal or unrelated to other canonical variates and each canonical variate is evaluated through the use of a chi-square test for significance.

METHODS AND MATERIALS

In this study, one set of variables represented the plant characteristics, i.e., plant height, length of the fourth internode, diameter of the fourth internode, and the number of stems per square meter. The fourth internode was chosen for measurement because it was the first fully developed internode. The second set of variables in the first analysis included: a measure of *Agasicles* feeding damage, i.e., percent of left area missing, the number of adult *Agasicles* per square meter, and the number of *Vogtia* per square meter. The location of the lakes and the number of samples taken from each of them were: Lake Alice, Gainesville, Fla., 76; Black Lake, Melrose, Fla., 155; Lake Lawne, Orlando, Fla., 34; Peples Pond, Garden's Corner, S.C., 2; Roadside ditch, Ft. Lauderdale, Fla., 12; and Twin Lakes, Summerville, S.C., 3. Samples were located by placing a 0.3-m metal frame on the alligator weed and the area within sampled. Sample sites were selected either by a random process (Lake Alice, Black Lake, Lake Lawne) or by taking samples at preselected, 5-m intervals (Gardens Corner, Ft. Lauderdale, Twin Lakes). In a second analysis, seven variables, representing the immature stages of *Vogtia,* replaced the variable which represented the total number of *Vogtia* per square meter. The deletion of the total number of *Vogtia* was necessary to comply with the orthogonal requirement of the program that no variable may repre-

*Research contribution of Dr. J. L. Brown, Entomologist, Florida Department of Health and Rehabilitation Services, Panama City, Florida, and N. R. Spencer, Research Entomologist, USDA, ARS, Gainesville, Florida.

TABLE 15.1

Impact Assessment of Alligator Weed on Water Resource Projects in the Southeastern States

	Degree of problem constraint		
	Extreme	Major	Moderate
Project objective			
Economic development			X
Environmental quality			X
Social well-being			X
Regional development			X
Project purposes			
Flood control	X		
Drainage	X		
Bank stabilization	X		
Water supply	X		
Irrigation	X		
Pollution control	X		
Wastewater management		X	
Sport fish and wildlife		X	
Recreation–boating		X	
Harbors and channels			X
Intracoastal waterways			X
Hydroelectric power			X
Project operations			
Planning and siting			X
Design and construction			X
Operation and maintenance			X
Flood plain management			X

sent a combination of other variables. A total of 287 samples from six lakes were used in the analysis as indicated in Table 15.2.

RESULTS AND DISCUSSION

In the first analysis, a set of plant characteristics was compared with a set of variables representing numbers of *Agasicles* and *Vogtia* present and a measure of *Agasicles* feeding damage. From these correlations, three new canonical variates were formed, as indicated in Table 15.3. Each of these new variates was significant at the 0.01 level of probability, indicating that three independent and significant linear relationships exist. If we examine the coefficients of the individual variables in Table 15.3 we can explain the combinations of variables making up each of these associations. The first canonical variable represents the most important of the linear associations, a positive association between plant height, internode length, internode diameter, and the number of *Agasicles* and *Vogtia*; it

represents a negative association between the same plant characteristics and *Agasicles* feeding. *Vogtia,* having a larger coefficient, favors these conditions more than *Agasicles*. The damage done by *Agasicles* places greater stress on the growing plants, which results in smaller, less thrifty stems.

The second canonical variate is of less importance than the first, but is significant. It is a negative association between *Agasicles* feeding and plant height, and internode diameter; therefore two relationships exist, one superior to the other. When plants are heavily attacked by *Agasicles,* the upper internodes may be damaged sufficiently to cause the plant to drop over and die.

The third canonical variate, independent of the first two and of less importance, consists of positive association between the number of *Agasicles* and internode length, and the number of stems. We conclude from this analysis that adult *Agasicles* tend to occur more frequently in dense stands of alligator weed with smaller stems, and this occurrence is negatively correlated with the number of *Vogtia* and with *Agasicles* feeding. The *Agasicles* feeding and stem size association is

TABLE 15.2

Variables used in Canonical Correlations for 287 Cases for Six Locations in North Carolina, Georgia, and Florida

Variable	Mean	Standard deviation
1. Plant	11.8955	3.9048
2. Internode length	22.2509	13.2195
3. Internode diameter	12.0070	5.7713
4. No. stems	51.2125	21.8647
5. Leaf area missing (%)	15.1742	26.3286
6. No. *Agasicles*	1.5366	2.6347
7. No. *Vogtia*	3.5157	4.1382
8. 1st instar *Vogtia*	1.2683	2.0959
9. 2nd instar *Vogtia*	0.6516	1.2192
10. 3rd instar *Vogtia*	0.4983	0.8766
11. 4th instar *Vogtia*	0.4564	0.9184
12. 5th instar *Vogtia*	0.4181	0.8807
13. *Vogtia* prepupae	0.0662	0.2881
14. *Vogtia* pupae	0.1777	0.4654

TABLE 15.3

Canonical Coefficients for Individual Variables, Analysis I

Variable	Canonical variate 1	Canonical variate 2	Canonical variate 3
1. Plant height	0.3455	−1.15221	−0.46995
2. Internode length	0.48798	−0.02244	0.62823
3. Internode diam.	0.32913	0.72620	0.18807
4. No. stem	−0.02748	−53516	−0.90613
5. Leaf area missing (%)	−0.63810	0.70406	−0.55411
6. No. *Agasicles*	0.32101	0.39437	0.97945
7. *Vogtia*	0.64560	0.44395	−0.70026

Canonical Variates in Analysis I

Number of canonical variate sets	Chi-square	Degrees of freedom
1	156.99149	12[a]
2	75.90101	6[a]
3	12.57912	2[a]

[a]Indicates significance at the 0.01 level of probability.

greement with the first canonical variable but much smaller in magnitude.

In the second analysis, plant height, internode diameter, and number of stems were compared with the number of *Agasicles, Agasicles* feeding damage, and each of the seven immature stages of *Vogtia* present. In this correlation, four new canonical variables were formed as indicated in Table 15.4. An inspection of the canonical coefficients for the individual variables reveals a positive association between stem diameter and internode length with the fifth instar *Vogtia*. This relationship, as explained in the first analysis, has been observed in the field in areas where alligator weed was under severe stress. During temporary stress periods the aerial stems may become small, with the stems making up the mat remaining large. Another relationship occurring in the first canonical variable is a negative association between *Agasicles* feeding and stem size. This relationship, also apparent in the first analysis, is a result of *Agasicles* damage producing stress in the alligator

TABLE 15.4

Canonical Coefficients for Individual Variables, Analysis II

Variable	Canonical variate 1	Canonical variate 2	Canonical variate 3	Canonical variate 4
1. Plant height	0.03523	1.15617	-0.53797	-0.44992
2. Internode length	0.55020	0.20074	0.70207	-1.21265
3. Internode diam.	0.38765	-0.56132	0.13546	1.71444
4. No. stems	-0.22032	0.57813	0.84154	0.28784
5. Leaf area missing (%)	-0.34916	-0.85504	-0.35962	-0.16349
6. No. *Agasicles*	0.27362	-0.19562	0.90137	0.41249
7. 1st instar *Vogtia*	0.18798	0.14124	0.14704	0.25493
8. 2nd instar *Vogtia*	0.02859	-0.11465	-0.39172	-0.19063
9. 3rd instar *Vogtia*	0.02138	-0.14565	-0.28131	-0.82188
10. 4th instar *Vogtia*	0.23763	-0.09089	-0.33282	-0.63356
11. 5th instar *Vogtia*	0.54452	-0.07560	-0.03484	-0.63356
12. *Vogtia* prepupae	0.05162	-0.14775	0.12094	-0.24329
13. *Vogtia* pupae	0.11968	-0.07671	-0.11247	-0.27877

Canonical Variates in Analysis II

Number of canonical variate sets	Chi-square	Degrees of freedom
1	195.18961	36[a]
2	94.64266	24[a]
3	23.28067	14
4	5.74528	

[a]Indicates significance at the 0.01 level of probability.

weed, which results in smaller, less thrifty plants. The second canonical variate shows a negative association between *Agasicles* feeding and plant height. As discussed earlier, reduced plant height with increased *Agasicles* damage is a direct result of *Agasicles* attack. Since the plant suffers more damage in the top few internodes, the more severely attacked plants are much more prone to breaking over, and to dropping these upper internodes.

REFERENCES – PART III

1. Brown, J. L., *Vogtia malloi,* A Newly Introduced Pyralid for the Control of Alligatorweed in the United States, Ph.D. thesis, University of Florida, Gainesville, 1973.
2. Correll, O. S. and Correll, H. B., Aquatic and Wetland Plants of the Southeastern U.S., U.S. Government Printing Office, Washington, D.C., 1972.
3. Cooley, W. W. and Lohnes, R. R., *Multivariate Analysis,* John Wiley & Sons, New York, 1970.
4. Earle, T. T., Riess, K., and Hidalgo, J., Tracer studies with alligatorweed using 2,4-DC[14], *Science,* 114, 695, 1951.
5. Hitchcock, A. E., Zimmerman, P. W., Kirkpatrick, H., Jr., and Earle, T. T., Growth and reproduction of water hyacinth and alligatorweed and their control by means of 2,4-D, *Contrib. Boyce Thompson Inst.,* 16, 91, 1950.
6. Maddox, D. M., Bionomics of an alligator weed flea beetle, *Agasicles* sp., in Argentina, *Ann. Entomol. Soc. Am.,* 61, 1299, 1968.
7. Maddox, D. M. and Resnik, M. E., Determination of host specificity of the alligatorweed flea beetle *Agasicles* n. sp. with radioisotopes, *J. Econ. Entomol.,* 62, 996, 1969.
8. Maddox, D. M., Andres, L. A., Hennessey, R. D., Blackburn, R. D., and Spencer, N. R., Insects to control alligatorweed; an invader of aquatic ecosystems in the United States, *BioScience,* 21, 985, 1971.
9. Selman, B. J. and Vogt, G. B., Lectotype designations in the South American genus *Agasicles* (Coleoptera: Chrysomelidae), with description of a new species important as a suppressant of alligatorweed, *Ann. Entomol. Soc. Am.,* 64, 1016, 1971.
10. Weldon, L. W., A Summary Review of Investigations on Alligatorweed and Its Control, U.S. Department of Agriculture, Agricultural Research Service, Crops Research Division, CR 33-60, September 1960.
11. Weldon, L. W. and Blackburn, R. D., Herbicidal treatment effect on carbohydrate levels of alligatorweed, *Weed Sci.,* 16, 66, 1968.
12. Wunderlick, W. E., Waterhyacinth control in Louisiana, *Hyacinth Control J.,* 3, 4, 1964.
13. Wunderlick, W. E., Practical suggestions for a large scale aquatic weed control project, *Hyacinth Control J.,* 5, 6, 1966.
14. Zeiger, C. F., Biological control of alligator weed with *Agasicles* n. sp. in Florida, *Hyacinth Control J.,* 6, 31, 1967.
15. Zurburg, F. W., Some problems with aquatic herbicides, *Hyacinth Control J.,* 7, 33, 1968.
16. Vogt, G. B., Exploration for natural enemies of alligatorweed and related plants in South America, in Aquatic Plant Control Program Technical Report No. 3, Biological Control of Alligatorweed, U.S. Army Corps of Engineers, Waterways Experiment Station, Vicksburg, Miss., 1973, p. B3.
17. Weldon, L. W., Blackburn, R. D., and Durden, W. C., Evaluation of *Agasicles* n. sp. for biological control of alligatorweed, in Aquatic Plant Control Program Technical Report No. 3, Biological Control of Alligatorweed, U.S. Army Corps of Engineers, Waterways Experiment Station, Vicksburg, Miss., 1973, D1.
18. Blackburn, R. D. and Durden, W. C., Integration of Biological and Chemical Control of Alligatorweed, in Proc. Research Planning Conference on Aquatic Plant Control Project, U.S. Army Corps of Engineers, Waterways Experiment Station, Vicksburg, Miss., 1972, C3.

REFERENCES CITED

Part IV
Research for Control of Water Hyacinth

Chapter 16

INTRODUCTION TO WATER HYACINTH*

A Brazilian, Karl F. P. Martius, was the first naturalist to describe the plant that has come to be known as "the world's worst weed" and the "million dollar weed," the floating aquatic plant commonly called the water hyacinth (*Eichhornia crassipes* [Mart.] Solms). The species is described in a work entitled *Nova Genera et Species,* which would indicate that the water hyacinth was probably not widespread in South America at the time Martius's work was published in 1824. From this presumed native habitat the water hyacinth has spread to most of the tropical and subtropical countries of the world.

DISTRIBUTION

Whether it was introduced accidentally to the United States by aquatic birds or deliberately by man is not known, but in those states where its existence matters most, Florida and Louisiana, the latter explanation is that accepted by authorities on the problem. In 1884 the water hyacinth was said to have been displayed at one of the exhibits at the New Orleans Cotton Exposition; visitors took samples of the beautifully flowering plant for their garden pools and fish ponds, and when within a short time the plant had so proliferated that the pond surfaces were covered with vegetation, unwanted plants were discarded in nearby streams, lakes, and bayous. The plant was unknown in the waterways of Florida before 1890. In that year, water hyacinths removed from a private pond and disposed of in the St. Johns River near Palatka created the initial infestation that was to spread over the main river, its tributaries, connecting creeks, and lakes in central and northeastern Florida, and, almost as rapidly as they were constructed, the drainage and irrigation canals of central and southern Florida. Today it is estimated that about 25,000 of an approximate 100,000 ha of fresh water in the many lakes, ponds, rivers, and canals of Florida are covered with water hyacinth. There are 500,000 ha of water in Louisiana in its 5,000 km of waterways and numerous inland lakes; estimates of water hyacinth coverage range from 50,000 to 100,000 ha. Alabama, Georgia, Mississippi, Texas, and

California waterways are also infested, but to a lesser degree (see Figure 16.1). These acreages would be considerably higher if it were not for control programs now in operation at a cost of more than $1,500,000 annually.[5,10,12]

The water hyacinth is a major problem in parts of Australia, in India, Ceylon, Java, and on the continent of Africa, where it was introduced to Egypt in the period 1889 to 1892, at about the same time it made its appearance in the United States.

Discovered in the Sudan in April of 1958, the water hyacinth within a few months had spread up and down 1,500 km of the White Nile; Sudan officials have authorized an annual expenditure of almost $1,500,000 for its control. Side tributaries also became infested, and from one of these the water hyacinth invaded the waters of Ethiopia, adding to the growing list of countries in which it has become established. Speaking of control efforts in Sudan, which parallel those practiced in the United States, an official of the Agency for International Development in Khartoum indicates that of major importance in the control program are efforts to contain the present area of infestation and prevent spread into the famous Gezira cotton irrigation scheme, the Blue Nile, and downstream in the Main Nile, north from Khartoum, toward the Aswan Dam. Penetration into any of these areas would be calamitous. The Egyptian government has recently sent botanists to this country to investigate methods of aquatic weed control in anticipation of possible infestation of the Aswan High Dam in 1975.

There is worldwide awareness of the potential danger of spreading aquatic weeds. A desire to avoid fresh infestations where possible, or at least to be able to cope with them at the earliest stages, has led to the formation of unofficial organizations to augment governmental research and control operations. Symposia bringing together interested scientists and officials are frequently held to correlate and disseminate information useful in controlling noxious plants.

The scientific advisory council for the Commission for Technical Cooperation in Africa South of the Sahara (CCTA) held a symposium in 1957 on

* Abridged and updated from Gangstad, E. O., Solymosy, S. L., and Zurburg, F. W., "Water Hyacinth Obstructions in the Waters of the South Atlantic States," House Document 37, U.S. Government Printing Office, Washington, D.C., 1957.

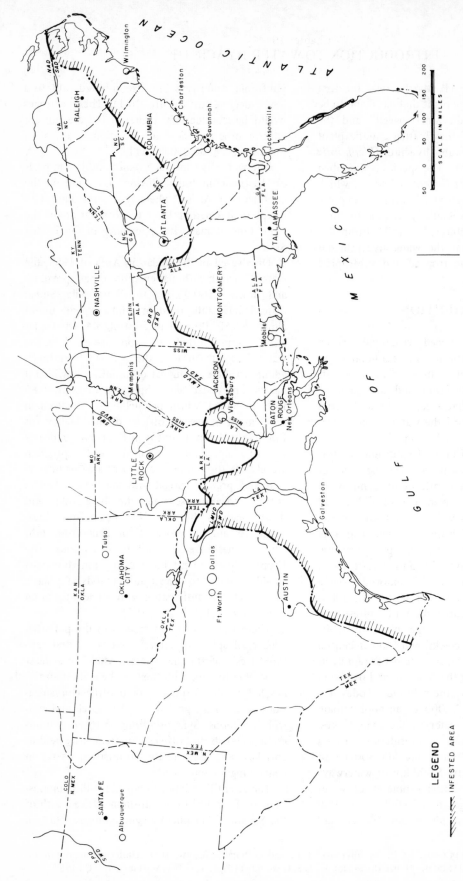

FIGURE 16.1. Map of southeastern states showing the area of infestation of water hyacinth (1965). (From House Document 251, 89th Congress, U.S. Government Printing Office, p. 36, 1965.)

the water hyacinth, which recommended that information helpful in identifying pest plants, or useful in describing potentially harmful plants that might not ordinarily be recognized as such, be widely distributed. In consequence, a pamphlet on identification of harmful aquatic plants was published in 1961. The hope is expressed in its preface that the pamphlet will make it easier "to give early warning of new infestation by harmful aquatic plants and thereby save time and money in their eradication."

NATURE OF PROBLEM

The immediately discernible effects of water surfaces being covered by water hyacinths to some extent can be expressed as quantitatively measurable economic losses, or, conversely, by estimates of economic benefits derived from removal of the growths. This is true mainly for agriculture, navigation, water management (flood control and irrigation systems), wildlife conservation, and commercial fishing, but also applies to the recreational activities of boating, swimming, sport fishing, and hunting. Interference with public health programs, which include control of mosquitoes and the maintenance of pure domestic water supplies, also occasions expenditures that can be directly attributable to these water weeds in some instances. Economic consequences of diseases causatively associated with the introduction of water hyacinth has not been calculated but is considered to be of serious proportions.

The long-term ecological effects of water hyacinth and alligator weed are not so readily apparent as are those mentioned above. The direction of natural successional processes tends toward the transformation of inland aquatic habitats into terrestrial ones unless some powerful physical force such as waves, streamflow, or wind interferes. Small lakes and ponds are not generally subjected to severe physical stresses of this kind, and, at the elevations characteristic of Florida and Louisiana, streamflow in most of the canals and rivers is visually almost imperceptible. The choking of these open water areas, as well as swamps, with vegetation can bring about a rapid shift from aquatic to edaphic, or terrestrial, ecosystems.

Marshlands in subclimax stages, rather than climax stages, provide the most suitable environment for wildlife, and marsh management practices intended to conserve wildlife are designed to augment, or in some cases duplicate, natural disturbances of drought, fire, floods, hurricanes, and animal overpopulation which maintain marshes at a subclimax or preedaphic level of succession. Water hyacinth and alligator weed interfere with this conservation practice of reversal of plant successions to the submersed aquatic and emergent stages; the introduced water hyacinth shades out desirable native submerged or floating freshwater aquatics, and extensive alligator weed growth eliminates wildlife food plants in fresh and brackish waters.[11-13]

In marshes, farm ponds, small tributaries, and even in larger bodies of open water, mats of water hyacinth provide a platform for terrestrial plants. The mat is transformed into a "floating prairie," or flotant, supporting plants such as cattail, giant bulrush, maiden cane, saw grass, and water fern. When sufficient sod is formed by an accumulation of windborne and waterborne soil and humus from decomposing plants, willows invade the flotant and the unusual situation of land being built from the water surface downward results in accelerated succession from aquatic to edaphic ecosystems and an unwanted loss of surface water areas.

Responsibility for maintaining the navigable waterways of the nation clear of obstructions lies with the U.S. Army Corps of Engineers, and it is this agency, in cooperation with stage organizations, that assumes the greater part of the burden of water hyacinth control operations, especially on larger rivers and canals.

Inland waterways of the southeastern and southern coastal areas form an extensive network vital to the transport of supplies to the areas served by them, and the marketing of important regional products, including fish, furs, lumber, oil, salt, sulfur, Spanish moss, cotton, sugarcane, corn, rice, and citrus fruits. When delay or stoppage of this traffic occurs because streams are choked with water hyacinth, the economic loss is considerable.[5,32-34]

IRRIGATION AND FLOOD CONTROL

The principal areas affected by water hyacinth and alligator weed are regions of heavy total annual rainfall. However, the precipitation pattern is irregular so that water management is necessary to prevent floods and to conserve impounded water for use during prolonged dry periods. Water for general human consumption, to irrigate crops, and to provide refuge for waterfowl, fish, and fur-bearing animals is made available as needed by systems of reservoirs and canals. In states such as Florida, where water hyacinths invade natural and artificially created freshwater habitats, control of these plants is a vital concern of water management agencies. Of particular concern in these important vegetable producing regions is the maintenance of water tables at proper levels during critical periods in the growing season. This is accomplished by regulation of water flow to and from water impoundments and natural lakes through canal and drainage systems.

In Florida, the Central and Southern Florida Flood Control District was created by the state legislature in 1949, after Congressional authorization in 1948. Total initial cost for the 2000 mi of canals and levees expected to be in operation at completion of construction, about 1968, was more than $380 million. The district is now spending approximately $140,000 a year on aquatic weed control; projecting current costs for chemical and mechanical control methods, it is predicted that their weed control program will cost as much as $500,000 a year when the entire project is complete.

Aquatic weeds impede the flow of water in canals and drainage ditches, in some instances reducing the flow by as much as 80%.[5] These weeds, and particularly the heavy mats of water hyacinths, decrease the amount of available fresh water because of their transpiration. Loss of water by this means is three to four times that lost in evaporation from open water surfaces. The destruction of water hyacinths may temporarily add to this loss, as studies indicate that respiration is increased when plants are injured by mechanical means or by toxic materials.[11]

Besides water management for the control of floods and the supplying of fresh water for crops and human consumption, the Flood Control District plans to include in its system of waterways barge canals for the transport of rocket engines to the Atlantic Missile Range at Cape Kennedy. Drainage on Cape Kennedy itself is important, and aquatic alligator weed has not yet been observed in these Cape drainage canals, although several submersed and emergent weeds are present.

WILDLIFE AND RECREATION

Marsh and water areas where food plants for fowl would normally grow are frequently taken over by water hyacinth and alligator weed, with the result that waterfowl are restricted to increasingly smaller acreages of water and a decreased supply of food. Overcrowding of valued waterfowl populations is considered a detriment to these species through potential increased exposure to parasites and disease, and because such concentrations expose them to more intensive destruction by hunters.

In 1948, the Fish and Wildlife Service estimated that in the southern states water hyacinths seriously interfered with utilization of resources that furnish recreation, food, and a livelihood to approximately 500,000 waterfowl hunters, sport and cane-pole fishermen, and trappers, and an estimated 7500 commercial fishermen. In the South Atlantic Division alone (states of North Carolina, South Carolina, Georgia, Florida, Alabama, and Mississippi) the fish and wildlife crop destroyed by water hyacinths is estimated at more than $4 million annually. An additional $873,000 annual loss is estimated from the presence of alligator weed in the streams of the division. Losses in Louisiana of the wildlife crop, which includes valuable fur-bearing animals, amounted to $14,727,000 in 1965.[5,13,34]

FISHING INDUSTRY

In a special scientific report issued in 1947. the Fish and Wildlife Service of the U.S. Department of the Interior analyzed damages to the fishing industry from aquatic weeds. Studies made in Louisiana and Florida considered the effects of aquatic weeds, especially the water hyacinth, the effects of control measures, and other factors which by themselves or in combination with water hyacinth infestation led to reduction or elimination of crop and game fish.

The Florida study paid particular attention to the fact that there had been an unexplained high mortality of fish in Everglades waters periodically before the water hyacinth became a serious problem. In order not to confuse the effects of the presence of water hyacinths and control methods used against them with unrelated factors, it was necessary to determine the reasons for these previous fish kills.

It was discovered that the muck soil of the Everglades contains a high concentration of hydrogen sulfide, particularly where the organic matter of which this soil is largely composed had once burned incompletely. Alkaline sulfides formed from the incomplete combustion of sulfur in the soil coming into contact with the slightly acid soil of wet, unburned muck, resulted in the formation of hydrogen sulfide, which, washed by rain into the waters of the canals and lakes, sometimes reached concentrations highly toxic to fish.

Hydrogen sulfide and carbon dioxide also accumulate in toxic amounts during the process of decomposition of plant material in the water; this was found to be one of the side effects resulting from control methods which kill the vegetation by either mechanical or chemical means, but it is most significant in the normal growth and decay processes of the vegetation — especially pronounced in northern areas where there is considerable winter dieback of the above-water parts of the plant. Chemicals presently used in control efforts are not toxic to aquatic fauna in the quantities used, and should not, therefore, in themselves contribute to an unfavorable environment for fish.

Besides the fact that killing the plant results in adding to the burden of toxic gases in the water, it was found that water hyacinths interfere with the aeration by which these gases would normally be dissipated and the oxygen supply renewed. In addition, the water hyacinth removes oxygen from the water for its own respiratory requirements. Reduced oxygen and increased carbon dioxide and hydrogen sulfide concentrations not only produce an unfavorable environment for fish, but also for plankton and aquatic insects on which the fish depend for food. Hydrogen sulfide apparently has no deleterious effect on the water hyacinth itself. Some animals considered undesirable for game or commercial fishing (gars, turtles, salamanders, snakes, and alligators) are able to live in this environment, preying on the remaining desirable fish populations (bass, shad, bream, catfish) that have been able to survive in these conditions. Dense growths of water hyacinths on the margins of lakes also smother spawning grounds of many game fish, and the scouring effect of drifting mats of water hyacinths destroys spawning beds and uproots desirable submerged vegetation.

Finally, when waterways become infested with the water hyacinth, commercial as well as game fishing is seriously hindered or eliminated depending on the density of the plants. Boats are unable to operate, and the use of various net and line-fishing devices becomes impossible.[5,11-13]

HEALTH

Water hyacinth is one of the many species of aquatic plants providing favorable breeding conditions for mosquito vectors of diseases. In the past, some forms of encephalitis and malaria have assumed epidemic proportions related to the density of carrier animals. The culicine genera *Culex* and *Aedes* are transmitters of equine encephalitis, yellow fever, and filariasis. An epidemic of St. Louis or equine encephalitis in Houston, Texas, first announced on August 20, 1964, is an illustration of the danger involved in permitting mosquitoes to breed. *Culex* mosquitoes have been identified as vectors in this outbreak, which had resulted in three proven deaths (and 19 suspected deaths) and a total of 77 proven stricken (and 479 suspected cases). City officials made insecticides available to the public at fire stations and enlisted the aid of Boy Scouts in a house-to-house campaign to advise the elimination of breeding habitats (even a water-filled tin can would be dangerous) and to urge precautions against mosquito bites.

The *Anopheles* genus is the sole known transmitter of causative protozoa infecting humans with malaria, the disease most common in the tropics where climatic conditions are favorable for the mosquito and for transmission of the disease throughout the year. While it is believed that there are probably more cases of malaria in the world than any other infection (it has been estimated that there still are approximately 3,000,000 malarial deaths yearly, with the entire population in some parts of the world infected more or less constantly), in the continental United States the disease is no longer a major threat. This is explained by public health authorities as the result of intensive mosquito control programs, coupled with a reduction of the animal reservoirs of the causative organisms, and radical prophylactic drug treatment which eradicates all stages of the parasite in human victims.

However, the biting habits of these insects (*Aedes molestus* was a serious pest in air raid shelters in London during World War II) makes them a continuing threat as potential transmitters of disease from infected animals to man, or from man to man. Mosquito control is, therefore, a major concern almost everywhere, and for those species which favor aquatic vegetation as a habitat the control of such vegetation becomes important also.

It is reported that the control of aquatic plants may be as important to prevention of disease transmission as direct control of these vectors. Eliminating the mosquito breeding habitat has proved less expensive than using larvicides in some cases. Relaxation of these efforts of aquatic plant control could produce epidemics of serious proportions.[12,13,25,36]

POTABLE WATER

Organic pollution in slow-moving streams and canals created by the growth and decomposition of water hyacinth and other aquatic vegetation is somewhat similar to that of sewage and industrial wastes; current control methods contribute to the problem of polluted water by creating masses of decomposing plants.

Occlusion of the water surface by plants reduces aeration, a natural means of purification as well as a means of renewal of the supply of dissolved oxygen in the water. It has been estimated that the oxygen-depleting pollutional load imposed by 0.4 hectare of growing water hyacinths is equivalent to the sewage from 40 people. Mats of water hyacinths clog the intake machinery of waterworks, thus increasing the cost of works operation. Tampa, Florida, in 1948, estimated that its added water treatment costs due to water hyacinths was from $5,000 to $10,000 a year. The added annual cost for the states in the South Atlantic Division for water treatment was estimated to be $50,000. Finally, as has been mentioned, the amount of available surface water for all purposes is reduced through transpiration of these plants.[5,12,13,34]

ECOLOGY, MORPHOLOGY, AND REPRODUCTION

The water hyacinth (*Eichhornia crassipes* [Mart.] Solms) is a monocotyledon belonging to the Ponterderiaceae or pickerelweed family. (The popular garden hyacinth, whose inflorescence may have suggested a resemblance, belongs to the Liliaceae family.) The plants vary in height from a few inches to 50 in. and form a bushy mass of fibrous roots 6 to 24 in. long. Perennial rhizomes 1 to 10 in. long are surmounted by a rosette of broad, glossy, dark-green petioled leaves, which are generally emersed. In small plants or in loosely connected stands, the petioles, containing spongy tissues, serve as bladderlike floats; in larger plants or in densely packed masses, the base of the petiole is not so pronouncedly bulbous. All parts of the plant except the seeds have specific gravity of less than 1.0, hence the plants are free-floating. The plant will also root in mud on the margins of lakes and in swampy areas. The inflorescence, consisting of from 2 to 38 flowers, is borne on a spike above the leaves. Individual flowers are about 1 to 1/2 in. in diameter, the lavender perianth having six lobes, with the banner petal displaying a chrome-yellow spot surrounded by a purple-blue border. Some insect pollination has been observed, but self-pollination when the flowers wither, which takes place from 24 to 48 hr after opening, is considered the most common means of producing ripe seeds. As the individual flowers wilt the spike twists and bends over; when the mature fruit capsule splits, the seeds are cast into the surrounding mat of hyacinths or into the water, where they sink to the bottom. The seeds remain viable for at least 7 years. Scarification (scratching the hard surface of the seed) by physical, chemical, or biotic means, and exposure to air, appear to be prerequisites to germination. A stand of medium-sized water hyacinths can produce as much as 45 million seeds per acre, but because relatively few of the seeds have the requisite conditions for germination only about 5% normally produce seedlings. Most of the seeds that do germinate are those washed to the water's edge and subsequently exposed when water levels are lowered, or those left in muddy areas by receding floodwaters. Loosened from the soil by rising water and later floods these seedlings can be carried into treated areas to renew infestations.

Reproduction is principally by vegetative means, daughter plants being produced by stolons which grow laterally below the water surface from the central rhizome, the interconnected plant forming enormous mats of vegetation. Various ways of describing the water hyacinth's prolificacy have been proposed, but one expression of this sufficient to make the point; plants can double in number in 10 days; in ideal conditions 100 plants will cover 1 km^2 in 8 months.[5,11-13]

TABLE 16.1

Eichhornia Crassipes (Mart.) Solms. (Syn. *Piaropus C.*) Deposited in the University of Southwestern Louisiana Ornamental Horticulture Herbarium

Sheet No.		Sheet No.	
08716	Louisiana, Evangeline Parish. Lake Chicot, around boat landing. Collection: S. L. Solymosy, No. 0618651, June 18, 1965.	08712	Louisiana, Evangeline Parish. Lake Chicot, around boat landing. Collection: S. L. Solymosy, No. 0618651, June 18, 1965.
08715	Louisiana, Lafayette Parish. In Wallis Pond, Lafayette. Collection: S. L. Solymosy, No. 0722601, July 22, 1960.	5342	Louisiana, Lafayette Parish. Wallis Pond. Collection: S. L. Solymosy, No. 0722601, July 22, 1961.
08714	Louisiana, St. Martin Parish. Martin Lake. Collection: S. L. Solymosy, No. 0112652, January 12, 1965.	04932	Photographs of specimen: Florida, Lake County. Vicinity of Eustis; introduced and well established in Lake Gracie. Collection: George V. Nash, No. 801, May 16, 1894. Courtesy of Royal Botanical Gardens, Kew.
08713	Louisiana, Lafayette Parish. In Wallis Pond, Lafayette. Collection: S. L. Solymosy, No. 0722601, July 22, 1960.	04932	Photograph of specimen: Mexico, Federal District. Valley of Mexico, floating on water. Collection: C. G. Pringle, No. 6316, June 2, 1897. Courtesy of Royal Botanical Gardens, Kew.
08711	Louisiana, Lafayette Parish. Old U.S.L. Barn, Lafayette. Collection: Eddie McWilliams, No. P395, September, 21, 1961.		

BOTANICAL SPECIMENS

Botanical characteristics have been studied in considerable detail representative specimens deposited in the herbarium of the University of Southwestern Louisiana are given in Table 16.1.

The plant is characterized by its bulbous petiole, shiny green leaves, mass of floating roots, and attractive lavender flower. It is a free-floating plant; the perenchyma tissue of the petioles provides buoyancy. Often a number of plants form a floating raft. The individual plants reproduce by offsets, and thus the mat can grow until it completely blocks the waterway; water hyacinth mats completely cover many of the water surfaces in some parts of southeastern United States.

Various attempts by chemical and mechanical means to control the weed have been undertaken in the United States. This plant, however, has an unusual ability to reinvade treated areas. Often, because of the plant's rapid rate of reproduction, it repeatedly clogs treated waterways.

Chapter 17

THE EFFECTS OF WATER QUALITY*

INTRODUCTION

Alligator weed and water hyacinth are two of the most troublesome aquatic weeds in the southeastern United States. Projects are presently being conducted by the Corps of Engineers which concern the biological control of these two species. During the course of these investigations, it has been apparent that not all bodies of water have the same problems with these weeds. Some sites have only alligator weed, some have only water hyacinth, a relatively few have both, and many have neither. Because the ranges overlap extensively, it has been assumed that temperature and solar insolation are not the factors separating them.

In order to determine the factors that separate the two weeds, two hypotheses have been formulated to evaluate the importance of water quality to the presence or absence of alligator weed and water hyacinth. The null hypothesis states that no difference in water quality is apparent among sites which have alligator weed, sites which have water hyacinth, and sites which have neither of these. If this hypothesis is accepted, it must be concluded that the presence of a species at a particular sites is entirely dependent upon its intrinsic ability to disperse, colonize, and compete. The alternate hypothesis states that some factor or factors of water quality separate the species and affect their ability to colonize and compete.[29]

To test these hypotheses, water samples were taken at various randomly chosen sites. Each site was classified H (water hyacinth present), A (alligator weed present) or N (neither of the two present). The water samples were analyzed for 15 variables, all of which (except pH) were converted to $\log_{10}$ to normalize the means. All calculations were made with the variables in log form. The normalized means (Table 17.1) were determined by converting the log values to antilog form. A multivariate discriminant analysis was then performed to determine which water quality factors were important in separating the three groups and which factors discriminate one from the others.[9]

PROCEDURES FOR ANALYSIS

The first procedure for analysis involved examining each variable to determine which one accounted for the most among-group variance. This was determined by calculating the F ratios

TABLE 17.1

Normalized Means and Standard Deviations[a] for Water Quality Parameters of Indicated Groups[35]

Variable	Water hyacinth	Alligator weed	Neither	Total
Alkalinity (total)	82.29 ± 1.48	100.84 ± 1.27	44.48 ± 1.46	66.09 ± 1.43
Calcium	19.28 ± 2.13	19.12 ± 2.89	13.61 ± 2.05	16.50 ± 2.24
Chloride	25.38 ± 2.11	46.24 ± 2.03	15.72 ± 1.64	24.05 ± 1.88
Copper	0.01 ± 0.05	0.01 ± 0.05	0.01 ± 0.04	0.01 ± 0.05
Fluorine	0.19 ± 0.22	0.26 ± 0.22	0.14 ± 0.24	0.18 ± 0.23
Hardness	79.26 ± 1.98	105.36 ± 1.41	48.70 ± 1.57	68.56 ± 1.69
Iron	0.25 ± 0.29	0.52 ± 0.65	0.19 ± 0.33	0.27 ± 0.39
Magnesium	2.97 ± 1.27	4.84 ± 1.51	2.02 ± 1.02	2.79 ± 1.20
Manganese	0.01 ± 0.02	0.01 ± 0.02	0.01 ± 0.03	0.01 ± 0.03
Nitrates and nitrites (total)	3.56 ± 0.41	3.38 ± 0.55	6.03 ± 0.50	4.12 ± 0.48
pH (−log [H+])	7.06 ± 0.84	7.13 ± 0.76	7.55 ± 1.06	7.29 ± 0.93
Phosphates (total)	0.71 ± 0.87	1.30 ± 1.28	0.40 ± 0.62	0.67 ± 0.85
Potassium	2.39 ± 0.68	2.90 ± 0.67	1.80 ± 1.02	3.20 ± 0.83
Sodium	15.65 ± 1.77	29.06 ± 1.37	8.80 ± 1.48	13.45 ± 1.56
Zinc	0.04 ± 0.14	0.01 ± 0.01	0.01 ± 0.01	0.02 ± 0.08

*By Center, T. D. and Balciunas, J., Entomologists, University of Florida, Technical Report 10, Appendix B, U.S. Army Waterways Experiment Station, Vicksburg Mississippi, 1975.

TABLE 17.2

Parameters Ranked in Order of Decreasing Importance in Discriminating Among Groups[3 5]

Rank	Parameter	F-statistic (df[a])
1	Sodium	11.56 (2/114)[b]
2	Total nitrates and nitrites	8.75 (2/113)[b]
3	Alkalinity	4.84 (2/112)[c]
4	pH	9.45 (2/111)[b]
5	Total phosphates	2.91 (2/110)
6	Iron	2.37 (2/109)
7	Manganese	1.73 (2/108)
8	Zinc	1.26 (2/107)
9	Potassium	0.94 (2/106)
10	Hardness	0.58 (2/105)
11	Magnesium	0.27 (2/104)
12	Calcium	0.85 (2/103)
13	Fluorine	0.15 (2/102)
14	Chloride	0.03 (2/101)
15	Copper	0.01 (2/100)

[a]Denotes degrees of freedom.
[b]Significant at 0.05 level.
[c]Significant at 0.01 level.

and choosing the parameter with the highest F value. This parameter was then deleted, and the procedure was repeated using the remaining variables. This continued until all parameters were ranked as to their importance (Table 17.2). Only four factors (sodium, total nitrates and nitrites, total alkalinity, and pH) proved to be significant. This procedure demonstrates which variables are important but does not demonstrate how they act in separating the groups.

The second procedure was employed to determine which variables affected each group. This involved the formation of coefficients for each parameter in a linear function (Table 17.3). The linear function (canonical variate) which accounted for the greatest variation was formed first. The most important variables in this function had the largest coefficient values. A second linear function was then formed which accounted for the greatest amount of the remaining variability. This procedure was continued until as much of the variability as possible had been accounted for. In this analysis only two linear functions were necessary. The first function accounted for 84% of the variability and separated the water hyacinth and alligator weed groups from the group with neither. Alkalinity was most important in pulling toward the weed groups and total nitrates and nitrites and

TABLE 17.3

Standardized Canonical Variate Coefficients Showing Contribution of Original Parameters to Canonical Variates[3 5]

	Canonical variate[a]	
	I	II
Percent of among-group variance	84	16
Cumulative percentage	84	100
Parameters		
Alkalinity (total)	1.157	0.482
Calcium	−0.339	0.394
Chloride	−0.078	0.148
Copper	0.000	−0.036
Fluoride	−0.087	0.120
Hardness	−0.353	0.089
Iron	−0.025	−0.648
Magnesium	0.337	−0.326
Manganese	−0.278	0.136
Total nitrates and nitrites	−0.581	−0.239
pH	−0.524	−0.709
Phosphates (total)	0.325	−0.302
Potassium	−0.041	0.459
Sodium	0.283	−0.933
Zinc	0.190	0.275

[a]Interpretations of canonical variates:
   a. Trophic state: High carbonate-bicarbonate alkalinity favors the presence of the weeds. This is correlated with calcium and hardness. All three factors are considered indicators of a eutrophic state. High hydroxide alkalinity (pH) and high nitrates and nitrites apparently favor the absence of either weed.
   b. Soluble salts: High soluble salts favor alligator weed over water hyacinth. Sodium is very important and is correlated with magnesium and chlorides. Iron and high pH are also important in discriminating the alligator weed sites.

pH in pulling toward the nonweed group. The second function accounts for the remaining 16% variability and separates the alligator weed from the water hyacinth group. Although no factors were specific for water hyacinth, sodium, iron, and pH were important in separating the alligator weed group.

To illustrate better how the parameters act within each linear function, the values for each coefficient may be plotted as vectors using each linear function as a new variable (canonical variate). The important variables are those that pull with the greatest magnitude in the direction of the respective groups (see Figure 17.1).

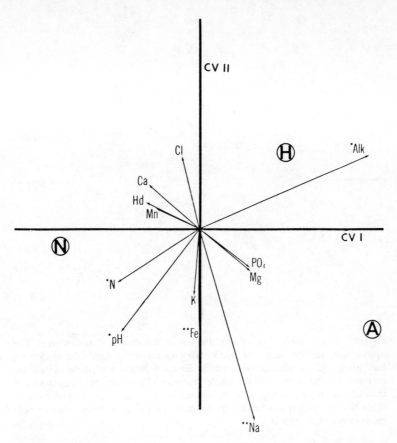

FIGURE 17.1. Graph of the canonical variate coefficients for each water quality parameter plotted as vectors on axes representing each canonical variate. Magnitude of the vectors represent the relative importance of the respective parameters. Orientation of each vector describes the direction of influence towards which a site is pulled if it has a high value for that parameter. The mean canonical variate coordinates for each group are designed by A, H, and N.

DISCUSSION OF RESULTS

The individual sites may then be plotted to determine the dispersion of the groups about these two axes (see Figures 17.2). By calculating the standard deviation about the mean along each axis and the slope for each group, ellipses representing the relative position of each group with regard to the canonical variates may be drawn. The area of overlap represents an area of ambiguity between the groups and may be interpreted as a zone of competition. The question that must be asked, however, is "Do these ellipses represent a preferendum for the respective species or are they indicative of a change caused by the presence of the plants?" A large stand of macrophytes will significantly increase the carbon dioxide concentration, model is not 100% accurate. If the total number

classified into each group is multiplied by the predictability for that group, a figure can be derived which represents the number of sites that should have been classified correctly. By dividing the number that were classified correctly by the number calculated to be classified correctly, a percentage is derived which represents the ability of each group to occupy its suitable habitat. Water hyacinth occupies 100% of the suitable sites, and 100% of the neither sites are unoccupied. Only 85% of the sites suitable for alligator weed were occupied by it.

It is felt that this reflects a greater ability of water hyacinth to disperse into the type of sites examined. Water hyacinth produces seeds which may be readily transported to these small bodies of water. Once the initial plants are present, they reproduce rapidly by vegetative propagation. Alli-

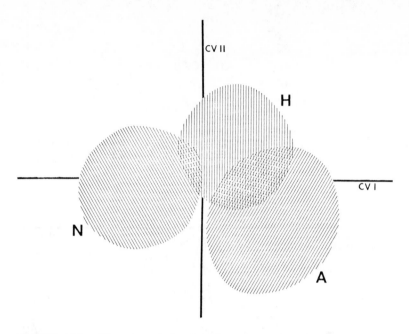

FIGURE 17.2. Dispersion of the sites about the canonical variate axes. The ellipses represent the mean canonical variate coordinates ±1 standard deviation for each group. If the parameters resulting in the separation of these ellipses are necessary for the presence of the species, then the ellipses represent dimensions of the realized niches or a zone of preferendum. In this case, however, it seems more likely that the differences in these parameters resulted from (rather than caused) the presence of the weeds.

gator weed, however, does not produce seed. It is dependent upon rafting or accidental introduction by man to be dispersed. Since these studies avoided river systems which would favor alligator weed dispersal, alligator weed was not found in some sites where conditions were suitable for it.

In order to determine the significance of the analysis in separating the group, an F matrix was formed (Table 17.5). This showed that the water hyacinth group differed significantly from the neither group, as did the alligator weed from the neither causing more calcium to go into solution as carbonates (higher alkalinity) and a decrease in pH. This may indicate why pH was higher for the nonweed group and alkalinity was higher for the weed groups. Likewise, the weeds would tend to draw nutrients out of the water and tie them up in biomass. This may explain why total nitrates and nitrites were higher in the nonweed group. Hence, the only real differences were probably caused by sodium and iron.

A site, then, may be classified as to group based on analyses of the water and where the point falls on the graph. By taking the original variables and classifying all the sites with the multivariate model

based only on their water chemistries, a classification matrix may be formed (Table 17.4). The percentage of each group classified correctly evaluates the predictability of the model. In this study, 70% of the sites were correctly classified based only on water quality. Sixty-four percent of the water hyacinth sites, 70% of the alligator weed sites, and 77% of the sites with neither were identified correctly.

By examining the classification matrix vertically, an index of suitability relative to the groups may be determined. For example, of the 41 sites predicted to be suitable for water hyacinth, 27 or 66%, were occupied by water hyacinth. Alligator weed occupied 57% of the suitable sites. Eight-two percent of the sites predicted to be unsuitable for the two weeds had neither species. However, it is known that the group. The water hyacinth and alligator weed groups did not significantly differ. Considering that the separation of the nonweed group was probably due to changes in the water caused by the plants, there is a tendency to accept the null hypothesis, that is, that there is no difference in water quality among the groups. However, as was pointed out pre-

TABLE 17.4

**Classification of Sites Based on Predictive Models
for Respective Groups[a]**

	No. sites classified into group			% classified correctly (predictability)
Group	Water hyacinth	Alligator weed	Neither	
Water hyacinth	27	8	7	64
Alligator weed	5	16	2	70
Neither	9	3	40	77
Percent classified correctly	66	57	82	70
(Suitability)	(103)	(85)	(106)	

[a]X^2 = 50.39, 4 degrees of freedom: significant at 0.005 level.

TABLE 17.5

**F Matrix for Testing Differences Between Groups (15/100
Degrees of Freedom)**

Group	Neither	Water hyacinth
Water hyacinth	3.92[a]	—
Alligator weed	4.58[a]	1.31[b]

[a]Significant at 0.01 level.
[b]Not significant.

viously, some separation does occur between the weed groups. Inclusion of other variables or an index of fitness for each site may accentuate this difference. From this analysis, two important factors can be determined. Alkalinity may be important in discriminating between the weed and nonweed groups, but it may also be due to the action of the plants. Sodium and iron are important in separating the alligator weed and water hyacinth groups. This does not mean that alligator weed occurs in saline conditions but merely that, relative to this model, sodium is generally higher at alligator weed sites.

Chapter 18
MECHANICAL METHODS OF CONTROL*

Mechanical control of aquatic vegetation may be defined as the utilization of mechanical devices to physically remove plants from, or destroy plants in, the aquatic environment. The type mechanical control system employed depends on the ultimate goal. Systems are available or may be developed for destroying plant material in the water and returning the debris to the water. Systems are available or may be developed for removing plant material from the water for disposal at a shore location. The latter system is generally referred to as aquatic plant harvesting. Such a system should prevent recycling of nutrients, build-up of organic control detrital material, increase of turbidity and/or depletion of oxygen. The operational feasibility of such systems remains to be established.

INTRODUCTION

A number of factors must be evaluated when determining the potential and requirements for mechanical harvesting. These factors include: (1) the type of plant to be harvested, (2) the type of water body, (3) the debris or other foreign matter encountered, (4) the nature of the shoreline, (5) the prevailing weather condition in the area, (6) the harvesting concept utilized, and (7) the plant disposal system contemplated.[36]

Type of Plant to be Harvested

The type of plant to be harvested is extremely important. Harvesting a floating plant, such as water hyacinth, requires a different approach than does a rooted submerged plant, such as hydrilla, or a rooted emerged plant such as cattail. Floating plants may be pushed to a centralized location and removed from the water. Rooted plants are fixed in the water body and the harvesting system must go to the plant (see Figures 18.1 and 18.2).

The plant's reproductive mechanism must also be considered. Hydrilla, as well as other plants, reproduce from extremely small fragmented sections; therefore, fragmentation must be controlled if spreading is to be prevented.

The rate of plant regeneration must also be considered when evaluating requirements for mechanical harvesting. Under ideal growing conditions, established stands of water hyacinth may double the area of infestation by vegetative means

FIGURE 18.1. Oil barges being towed through water hyacinth-jammed intercoastal waterway in Louisiana, 1945. (From House Document 37, 85th Congress, U.S. Government Printing Office, Washington, D.C., p. 226, 1957.)

*Research contribution of Larry M. Curtis, Mechanical Engineer, formerly of the Department of Natural Resources, Tallahassee, Florida, and P. A. Frank, Plant Physiologist, USDA, ARS, Davis, California.

FIGURE 18.2. Pasture land flooding caused by water hyacinth-choked canals near Raceland, Louisiana, 1945. (From House Document 37, 85th Congress, U.S. Government Printing Office, Washington, D.C., p. 222, 1957.)

in a 2- to 4-week period. Therefore, a harvesting system could be developed for eliminating an established growth of water hyacinth and a maintenance program could be established for constant surveillance and harvesting of newly generated plants.

Hydrilla and other rooted aquatics present a different problem from a regrowth standpoint. Most harvesters are designed for mowing rooted aquatics at some depth below the surface. This is similar to maintenance of a lawn. Previous experience has indicated that under ideal conditions hydrilla will grow back to the surface within 4 to 6 weeks after mowing 5 ft below the surface.

The quantity of plant material produced must also be considered. A total coverage of water hyacinth on a hectare plot may weigh 25 to 100t, depending on the size and density of the plants. The productivity of water hyacinth per hectare per year has been estimated at between 10 and 30 t metric tons of dry matter. Since water hyacinth contains approximately 95% water, a harvester system would have to remove hundreds of tons of wet plant material per hectare per year in a highly productive water body. Hydrilla is less bulky than water hyacinth and the tonnage per hectare is also less. This plant contains approximatley 90% water with an estimated productivity of 3 tons of dry matter per acre per year, or 30 t of wet material per hectare per year.

Type of Water Body

The second factor to consider is the type of water body. For example, a river or meandering stream would require different harvesting techniques than would a lake or a reservoir. Currents must be considered because of the effect they have on movement of floating plants as well as on maneuverability of harvesting craft. The physical dimensions of the water body would likewise determine the selection of transport and disposal equipment.

Debris or Foreign Matter Encountered

Closely related to the type of water body is the debris or foreign matter in the water body. In natural lakes or meandering streams, debris, stumps, logs, and other material foul or interfere with harvesting efforts. If harvesting is contemplated in a reservoir, this type material should be removed to minimize maintenance and to increase the efficiency of the operation.

Nature of the Shoreline

Shoreline characteristics play an important role in the efficiency of a harvesting operation. In natural situations aquatic growth is usually prolific along shallow shorelines. Growth of many aquatic species begins in shallow areas and spreads from there into deeper areas. For these reasons, it is desirable to have a deep shoreline with a minimum of growth. All trees and other vegetation should be removed and the shoreline deepened prior to construction of a reservoir where harvesting is contemplated.

Prevailing Weather Conditions

The prevailing weather conditions must be considered when designing a mechanical harvesting system. For example, water hyacinths are often floated to a fixed site for removal. Such a site should be located to take advantage of the prevailing wind. Likewise, wind, rain, and other inclement weather conditions will limit the number of days when a harvesting operation can be carried on.

HARVESTING CONCEPTS

A number of types of equipment are utilized for removing aquatic vegetation. This equipment ranges from commercially available devices such as draglines and cranes to equipment designed especially for harvesting aquatic vegetation. Generally, two harvesting concepts are employed with this equipment.

One concept is to cover the infested areas of a water body in a systematic manner with a harvester. Such a harvester normally must have a storage capacity as well as a means for transporting and unloading at a shore location. In Florida such systems have presented enormous problems because of the huge tonnages and bulk of harvested plants.

The second concept is to operate a harvester from a shore location. This system is normally used for removal of floating aquatics such as water hyacinth, and for removal of rooted aquatics in canals, and has the advantage of eliminating the need for on-board storage or transport craft. With this system support craft may be needed to break or cut plants free and to push plants to the harvester site. If this technique is used in a reservoir, the sites should be properly planned and prepared prior to construction.

DISPOSAL SYSTEM

A disposal system is required for plants that have been harvested and delivered to shore. At the present time, most harvested vegetation is hauled by truck to a convenient dumping site. In most cases plants are chopped prior to transport to

reduce the bulk and increase the carrying capacity of the trucks. For large-scale harvesting where huge quantities of plants are removed, this type disposal system would be unsatisfactory.

There are several possible alternatives to disposal. For example, large stockpiling conveyers might be used or the plants could be dewatered prior to disposal. The most attractive solution would be the development of a commercial product from the plants. This would allow processing equipment to be installed on site and then the product could be moved into commercial channels.[4,5,32,33]

MECHANICAL EQUIPMENT

Machines developed by the Corps included the following:

1. The U.S.S. Kenny: a larger crusher boat that removed the plants, pulverized them, and returned the refuge to the water (see Figure 18.3)
2. Sawboats: lightweight destroyer craft that killed water hyacinth by passing a gang of rotating saws over the plants until the plants were chopped sufficiently to kill them (see Figure 18.4)
3. Conveyer harvesters: harvesters that used conveyers to remove plants from the water and place them on shore
4. Floating derricks: derricks mounted on barges that were equipped with a grappling device and used to remove heavy growths from narrow streams and place the material on adjacent banks
5. Containment structures: permanent control structures or traps constructed at the lower end of large lakes on either side of navigable channels and temporary traps constructed at the entrance of small streams or across lagoons to prevent hyacinth from drifting into navigable streams. Small destroyer boats were used for mechanical destruction of trapped water hyacinth and small conveyers were used to remove the plants.

COMMERCIAL EQUIPMENT

Mechanical harvesting equipment is available in various sizes and configurations. Some of these are described below

Draglines
Commercially available draglines are used ex-

FIGURE 18.3. Water hyacinth destruction by a conveyer barge operating in Bayou Lafourche near Thiobodaux, Louisiana, in 1946. (From House Document 37, 85th Congress, U.S. Government Printing Office, Washington, D.C., p. 288, 1957.)

FIGURE 18.4. Mechanical destroyer barge cutting through a channel in Bayou St. John near New Orleans, Louisiana, in 1947. (From House Document 37, 85th Congress, U.S. Government Printing Office, Washington, D.C., p. 240, 1957.)

tensively for control of aquatic plants. They are most frequently used in canal maintenance operations where there is a dual goal of weed control and silt removal. For removing rooted plants, the dragline is moved along the bank of a canal and vegetation and/or silt is removed and deposited on a bank. Draglines may also be placed

at strategic locations such as bridge crossings t⟨ remove floating plants (particularly wate hyacinth) as they collect behind such obstruction⟨

For removing rooted plants, conventional dra buckets are used. Floating plants are mor efficiently removed using a hyacinth bucket, cla⟨ shell, or other type of bucket capable of grasping large mass of the plants. This technique is also used in lakes for removing water hyacinth that has floated to a collection point.

"A" Frame Drag Boats

The Central and Southern Florida Flood Control District, as well as Dade County's Public Works Department (Florida), use boats to tow an "A" frame device along the bottom of canals for weed control. The boats are designed especially for this job and have sufficient horsepower to readily tow the A frames. The A frames cover an approximate 8-ft width. Their purpose is to tear hydrilla free from the bottom. The freed plants are then pushed to a collection point where a dragline or crane is used for removal.

The A frame has the disadvantage of increasing turbidity due to agitation of canal bottoms. This technique also rapidly spreads vegetatively reproductive plants such as hydrilla by breaking and releasing plant fragments, which then spread up and down the canal. It should not be used in areas where infestation covers only a portion of the water body, as it is desirable to avoid spreading the infestation.

Dredges

Conventional dredges are not normally used for aquatic weed control. They do offer some control by ingesting rooted aquatics along with sand, muck, and other dredge material. One dredge device that has been used to some extent in Florida is the "Mudcat." This machine is designed to remove up to 18 in. of the upper layer of hydrosoil and can remove the material from a maximum depth of 3 m. This device uses a horizontal auger to remove the hydrosoil over an 8-ft-wide front. This material is then pumped through a discharge line up to 2500 ft to a spoil basin.

Conveyer Harvesters

Machines using conveyer systems to handle aquatic plants are referred to as conveyer harvesters. Different techniques are used for such systems.

Mobile harvesters are available that cut rooted plant material 2 m below the surface, elevate this material into a storage hopper, and use shuttle craft to transport the plant material to shore. These harvesters are also capable of harvesting floating plants. Typically these harvesters have an 8- to 10-ft-wide pick-up conveyer to elevate the plants from the water.

Conveyer harvesters are designed specifically for water hyacinth. These harvesters are typically 10-ft wide shore-based units. Wind, water currents, pusher boats, and booms are used to move the plant material to the shore site. The Florida Department of Natural Resources has developed a large 13-m-wide fixed site conveyer harvester for removing water hyacinth on the St. Johns River. Fixed-site water hyacinth harvesters typically have three conveyer systems: a pick-up conveyer to lift the plants from the water, a cross conveyer for moving the plants to shore, and an elevating conveyer for loading on a vehicle. Some are equipped with choppers for reducing the bulk of the plants. Only in very specialized cases where small acreages are involved have conveyer harvesters provided total aquatic weed control.

Cutter Boats

A number of companies manufacture cutter devices to mow a swath through rooted aquatics. These devices may be self-contained or may attach to the front of a fishing boat. Such cutter boats are practical for clearing trails or boat paths through thick rooted plants, but are not practical from an aquatic plant removal standpoint.

DOCUMENTATION OF HARVESTING

Reliable data concerning the economics of harvesting aquatic plants are minimal but some information concerning water hyacinth harvesting efforts are given in Table 18.1.

Shell Creek Project

This project evaluated a conveyer type fixed-site water hyacinth harvester constructed by Sarasota Weed and Feed, Inc., Sarasota, Florida. The machine was designed specifically for harvesting water hyacinths. The machine was operated for 64 days during the summer of 1971. The machine was leased by the state and operated by the Game and Fresh Water Fish Commission and the Corps of Engineers. Data were collected relative to the time and motion aspects of the operation, costs incurred, plant characteristics, and water quality.

Gant Lake Canal Project

This project evaluated the conveyer type fixed-site water hyacinth harvester previously tested at

TABLE 18.1

Data on Three Harvesting Projects

	Harvesting sites		
Volume comparison	Shell Creek	Gant Lake	Trout Lake
Days operated	21	23	26
Hours operated	155.2	151.36	208.0
Hectares harvested	8.26	3.42	5.68
Metric tons harvested	2,326	1,460	1,470
Metric tons per hectare	50	75	62
Cost comparison	Footnote a	Footnote a	Footnote b
Total cost	$32,781	$3,451	$6,603
Cost per hectare	3,969	1,009	1,065
Cost per metric ton	13.82	2.32	4.79

[a]Includes cost for disposal vehicles.
[b]Excludes cost for disposal vehicles.

Data from an unpublished cooperative research study of the Department of Natural Resources and the Florida Freshwater Game and Fish Commission, Tallahassee, Florida, and the U.S. Army Corps of Engineers, Jacksonville District, Jacksonville, Florida, 1971–1972.

Shell Creek Reservoir. The harvester was modified following the effort at Shell Creek. The machine was operated by Sarasota Weed and Feed, Inc., and data were collected by Game and Fresh Water Fish Commission personnel for a 23-day period during the winter of 1972. Data were collected relative to time and motion aspects, plant characteristics, and costs incurred.

The data collected in this test are probably the most realistic to date because the plants were removed from a completely covered canal where more exact measurements of area, plant density, etc., were possible.

Trout Lake Project

This project evaluated a prototype fixed-site water hyacinth harvester owned by Linder Industrial Machinery Company, Inc., Lakeland, Florida. This machine was operated by Game and Fresh Water Fish Commission personnel with support from Lake County for a 26-day period during the summer of 1972. Data were collected relative to the time and motion aspects of the operation, plant characteristics, and costs incurred.

ROOTED AQUATICS

Little reliable cost data exist for harvesting rooted aquatics. One of the more extensive harvesting efforts in Florida was conducted between 1966 and 1969 by the city of Winter Park in Orange County. Winter Park conducted a 4-year program in an attempt to control hydrilla on approximately 1000 acres of lake property within the city. Although some cost figures are available, significant weed control was never obtained and the harvesting project was finally abandoned. A total of $128,599 was invested by Winter Park in purchase of harvesting and support equipment during this period.[4]

An effort that has been successful is the hydrilla control program conducted by the Central and Southern Florida Flood Control District. This program utilizes the tugboat A frame concept mentioned previously. Although this control method is not considered the ideal solution here, they are able to keep canals open by use of constant maintenance program. The costs per acre average less than $40, which is far below other harvesting efforts.

UTILIZATION RESEARCH

Livestock Feed

The palatability to livestock of feeds processed from water hyacinth compares poorly with that of most other normal feeds and forages. Animals can be induced to eat the processed water hyacinth if mixed with molasses or other feeds, but even in these instances, if the proportion of water hyacinth in the feed is too high, intake is reduced and the animals lose weight. As little as 5% water hyacinth in the diets of hogs depressed weight gains. In diets containing 30% water hyacinth, weight gains were reduced 94%. Feeding tests with ruminants were more favorable.[2]

Ensiled Water Hyacinth

One means of eliminating the expense of drying aquatic plants in the preparation of livestock feeds is to convert the plant material to silage. The production of silage from water hyacinth has been studied fairly extensively by workers at the University of Florida.[2] These studies can be summarized as follows:

1. Acceptable silage can be produced from water hyacinth by mechanically removing 50% or more of the plant moisture and by adding a source of free carbohydrate.
2. From 2 to 4% dried citrus pulp or cracked corn were suitable sources of free carbohydrate.
3. The quantity of silage consumed by cattle increased as the level of added carbohydrate increased.
4. The addition of sugarcane molasses alone did not improve the acceptability of silage by cattle.
5. Water hyacinth alone did not produce a usable silage.

Soil Amendment

The water hyacinth plants from numerous locations in Florida were analyzed for composition of nutrients important in plant growth. The average levels of major nutrient elements were considered adequate for those materials to be used as soil amendments. The average C/N ratio (23:1) compared favorably with the range of C/N ratios (20:1 to 30:1) for legumes and was much better than that (90:1) of most straws. Ground, dried water hyacinth was added to a number of different virgin Florida soils at rates of approximately 5.5, 11, and 22 t of dry matter per hectare. The results obtained were precisely what one would have expected from the addition of equivalent quantities of similar organic matter to the soils. Consequently, we can infer from this work that water hyacinth is suitable as a soil amendment. The disadvantage in using water hyacinth for this purpose is the need to transport about 70 t of fresh water hyacinth per hectare to achieve the lower level of addition of dry matter and about 280 t for the highest.[18]

Source of Paper Pulp

A major proposal for utilization of aquatic weeds such as water hyacinth is its use as a possible source of pulp for making paper. Extensive testing of water hyacinth[14] for this purpose was completed recently at the University of Florida Pulp and Paper Laboratory. They were not found satisfactory. Reasons given for the failure of the water hyacinth to produce suitable pulp were

1. Very low fiber yield (13%)
2. Excessive shrinkage of the paper on drying, resulting in wrinkling
3. Brittleness of the paper
4. Paper turning very dark on drying
5. Poor tear properties

Chapter 19

PROGRAM OF CHEMICAL CONTROL*

The original federal project in the United States authorized the operations of a program for removal of water hyacinth in the navigable waters so far as they become an obstruction to commerce and related water resource activites. This program was extended nationwide under the 1965 authority (P.L. 89-298; 33 U.S.C. Section 610) to include water hyacinth, alligator weed, Eurasian water milfoil, and other obnoxious aquatic plants. Careful study of the use of herbicides for control of aquatic plants indicates that detrimental effects on the environment are short-lived and can be largely avoided by proper selection of herbicide and method of application.

INTRODUCTION

The quality of the environment has been a major concern of the Corps of Engineers as part of the basic mission to improve river and harbor transportation, to provide more adequate flood control, and to plan for appropriate water resource development. Because aquatic plants do interfere with navigation, flood control, and water resource development, the Corps of Engineers has frequently been involved in the control of aquatic plants. While mechanical methods were first used for this purpose, chemical methods have since been developed for most situations. There are, however, important limitations to chemical control, and these limitations must be properly dealt with for protection of the environment.

The most effective water hyacinth killing chemical yet discovered is 2,4-dichlorophenoxycetic acid, commonly caled 2,4-D. This herbicide is an organic chemical belonging to a group of substances known as plant hormones or growth regulators. It has been recommended widely for aquatic plant control and continues to be the most frequently used herbicide. It is noncorrosive, nonflammable, and nonirritating to the skin. It has been shown to be nontoxic to humans and animals when ingested to concentrations that are expected to follow its application as a herbicide. Its toxic ffect on fish has been shown to be negligible. any crops and ornamental plants or shrubs are dversely affected by 2,4-D, but coordination of a spraying program with the season, when crop production is at a minimum, is helpful in preventing injury due to accidental drifting of the spray. Proper care must be exercised in spraying operations because careless handling of spray apparatus could lead to serious crop losses through drifting of the spray into fields. The possibility of losses through contamination of irrigation water is remote because of the dilution factor. In general, depending upon size and depth of the body of water, and whether it is freely flowing or sluggish, there are few or no harmful effects from hyacinth control operations on water supplies. In a few instances where the supply source is very heavily infested and where there is little discharge, special precautions need to be taken by treating portions of a body of water. Practically no adverse effects on water supplies have been reported since the widespread use of 2,4-D for hyacinth control operations was started by the Corps and state and local agencies.

In addition to the 2,4-D acid itself, a number of its derivatives are available in several formulations, including amine, sodium, and ammonium salts and some alkyl and aryl esters. Commercial preparations in liquid, powder, dust, and pellet form are on the market in water-soluble and oil-soluble forms in a range of concentrations. Proprietary formulations contain varying amounts of 2,4-D acid equivalent. In order to provide a standard scale for expressing the rate of application per acre, the usage is expressed in terms of pounds of 2,4-D acid equivalent. The amine salt of 2,4-D, which is the least volatile, and which can be mixed with water for the carrying agent, is used in our program for water hyacinth control. Good results are obtained using this type mixture when applying the 2,4-D at an average rate of 2 lb/acre (2.24 kg/ha). Higher rates of treatment, between 2 lb/acre and 4 lb/acre (2.24 kg/ha and 4.48 kg/ha) may be required, depending upon the size of the plants, the density of infestations, and weather conditions.

METHODS OF APPLICATIONS

The proper amount of chemical concentrate,

*Adapted and revised from "Expanded Project for Aquatic Plant Control," House Document 251, U.S. Government Printing Office, Washington, D.C., 1965.

depending on the acid equivalent per gallon, is mixed with 50 gal (190 l) of water in a 55-gal (209 l) steel drum connected to a pump and applied directly as a spray. The concentrate can also be placed in a tank metered to a source of water where it becomes mixed at proper porportions as it is used. The 2,4-D can be applied in as little as 50 gal of water per acre (46.75 kg/ha) but may be applied in mixtures from 100 gal to 200 gal of water per acre (93.50 kg/ha to 187.00 kg/ha). The ester salt of 2,4-D mixed with a light fuel oil is particularly adaptable for treatment by planes when drift is held to a minimum. The spray equipment used consists of Bean® spray pumps of the 10-gal/min (38 l/min) and 15-gal/min (57 l/min) models driven by suitable air-cooled gasoline engines. The concentrate is mixed-in-line with water from overboard. The gun used is made from pipe fittings and copper tube fittings with a suitable quick-action shut-off valve. It is swivel-mounted on a stanchion to absorb the recoil. Pressures of up to 400 psi (2,760 kg/m^2) are used to penetrate dense vegetation. Less pressure is used in proximity to valuable vegetation to reduce the chance of damage from drifting chemicals. Where extreme infestations occur in small impoundments and where there is a hazard of deoxygenation, a special technique is used. The procedure essentially includes one or more of the following: (1) Obnoxious growths are treated at or near feeder areas; (2) growths are trapped in small quantities and treated; (3) retreatment is made about 8 weeks following the first application of herbicide; (4) only 25% to 50% of infestations are treated at one time; and (5) dead plants are permitted to sink before treating adjacent live plants.

FORMULATION AND RATE

Experimental studies on herbicide formulations were conducted by the U.S. Department of Agriculture near Fort Lauderdale, Florida, 1963, with the following objectives: (1) to test various formulations of 2,4-D and compare their relative toxicity to water hyacinth, (2) to determine the optimum rate and gallonage at which herbicides should be applied, (3) to determine the effect upon plant kill of adding certain wetting agents to the spray solution, and (4) to compare the herbicidal responses of water hyacinth. All herbicides tested were effective with only minor differences due to formulations applied at 2.24 to 4.48 kg/ha.

Accordingly, it has been the policy of the Corps of Engineers to stay within these rates of application (see Table 19.1).

OPERATIONS COST

Costs of hyacinth control operations vary with the amount and density of infestation, with the type of equipment used, with the character of access to the area, and often with weather conditions. Experience indicates that costs range from $10 to $75/acre ($4.05 to $30.37/ha) for each application of spray herbicide. The lowest unit costs are obtained when large areas are treated by aircraft and the highest when areas to be treated are remote and access is difficult.

Costs of water hyacinth control on Lake Blackshear by the Crisp County Power Commission under the expanded project are summarized as follows:

Field costs – 523 acres at $23,22 per acre	$12,146
Supervision and administration	$2,873
Total	$15,019

Costs of a similar operation in Louisiana, including supervision and overhead charges, were approximately $40.00/acre ($16.20/ha) for treatment of the vegetation in an area difficult to reach. Satisfactory results have been obtained with an application of not less than 2 lb acid equivalent per acre of the amine salt of 2,4-D. To cover borderline cases, however, the New Orleans District uses approximately 4 lb acid equivalent per acre. The cost of the concentrated chemical purchased in 80-drum lots averages about $4.00/gal of the 4 lb/gal material ($1.05/l). These costs have increased during the current program period, but are within the range of normal change due to inflation.

ENVIRONMENTAL CONSIDERATIONS

Direct effects of repeated use of herbicides such as 2,4-D on the aquatic environment are relatively minor if the herbicides are used properly. Indirect effects may conceivably result in particular changes in the environmental system over a period of time. These changes, however, are not necessarily detrimental, for single organisms are not indispensable to an ecological community and many different organisms compete for the same

TABLE 19.1

TABLE 19.1

Effects of Phenoxy Herbicides on Topkill and Regrowth of Water Hyacinth Applied at 2.24 and 4.48 kg/ha

Treatment chemical	Rate (kg/ha)	Months after application	
		1 (topkill) (%)	3 (regrowth) (%)
2,4-D ethyl ester	2.24	96	8
	4.48	96	7
2,4-D butyl ester	2.24	93	13
	4.48	96	5
2,4-D isopropyl ester	2.24	90	16
	4.48	96	8
2,4-D BE ester	2.24	92	15
	4.48	96	8
2,4-D isoctyl	2.24	93	9
	4.48	97	2
2,4-D alkanolamine	2.24	77	22
	4.48	90	8
2,4-D dimethylamine	2.24	89	13
	4.48	93	4
Silvex®	2.24	96	5
	4.48	98	2
MCPA ([4-chloro-o-toloxy] acetic acid)	2.24	87	15
	4.48	94	8
2,4-DP	2.24	87	15
	4.48	94	8
Untreated check	—	1	1
	—	1	1
Significant difference (P = 0.05)		13	9

Research data from a cooperative study of the U.S. Army Corps of Engineers, the U.S. Department of Agriculture, and the Florida Agricultural Experiment Station; Annual Report, Fort Lauderdale, Florida, 1963.

iche. The most obvious ecological effect of herbicide application is the reduction of the plant population to an earlier point in the succession. This again is not a serious or lasting problem because many different organisms are known to be able to adapt their metabolic processes to utilize the herbicide as a sole source of carbon and reduce it to a nontoxic entity.

The concentration of detectable amounts of the herbicide 2,4-D in surface water is known to be determined by such abiotic (nonliving) factors as the rate of treatment and the extent of dilution as affected by depth, flow, or addition of water, or both, to the system. Photodegradation and absorption of the hydrosoil are not found to be major routes for reduction of 2,4-D residues, but biodegradation by microorganisms and metabolism by plants are found to be major routes of herbicide decomposition. Temperature and time are important parameters affecting the rate of dissipation because they are known to influence biological processes to a large degree.

Chapter 20
RESIDUE STUDIES IN SMALL PONDS*

A field study was conducted to determine the uptake and dissipation of the dimethylamine salt of (2.4-dichlorophenoxy) acetic acid (2,4-D) in water, hydrosoil, and fish in ponds located in Florida and Georgia. The residue in the treated water was dissipated to less than the negligible level of 100 ppb in 2 weeks, and did not attain this level at any time in the hydrosoil or fish flesh, from a treatment of 2.24 to 8.96 kg acid equivalent per hectare of 2,4-D.

INTRODUCTION

The widespread occurrence and uncontrolled growth of various aquatic plants, especially in the southeastern United States, has caused many problems including interference with navigation, obstructed water flow, lowered real estate values, reduced fishing success, and impaired recreational use. In addition, the water hyacinth promotes water stagnation and additional breeding areas for mosquitos.

Various methods have been used to control aquatic plants, including mechanical harvesting, which though feasible in some areas, is too costly and gives only short-term results. At present, no effective biological control method is operational for the most troublesome species, namely, water hyacinth. However, water hyacinth can be controlled by the use of phenoxy herbicides. The effects of these herbicides on the aquatic environment have been studied but have not been fully examined for cause and effect in the field.[2,4,11-13]

This study is a compilation of data from experiments designed to determine the residue levels and rate of dissipation of the dimethylamine salt of 2,4-D in water, hydrosoil, and fish from ponds treated at 2.24, 4.48, or 8.96 kg 2,4-D acid equivalent per hectare (2, 4, or 8 lb acid equivalent per acre). These applications are one half, one, and two times the recommended treatment rate. Ponds were located in two widely separated geographical locations to study the effects that different physical and chemical characteristics of the aquatic environment might have on the uptake and dissipation of the herbicide. The results of these

studies of ponds in Florida and Georgia and the summary of treatment means for rates of applications and days after treatment are given in Table 20.1.

POND STUDIES IN FLORIDA

Three ponds were located on the Plantation Golf Course at Crystal River, Florida. These ponds serve only as water hazards on the golf course, and are not used for drinking or irrigation purposes. All three ponds were treated with rotenone in May 1971 and restocked in June 1971 with large-mouth bass (*Micropterus salmoides* Lacepede), channel catfish (*Ictalurus punctatus* Rafinesque), bluegill (*Lepomis macrochirus* Rafinesque), and redear sunfish (*Lepomis microlophus* Gunther). One month prior to treatment the three ponds were stocked with water hyacinth to cover to 10% of the surface area at the time of treatment.

POND STUDIES IN GEORGIA

Four private ponds within a 16.1-km radius of Warm Springs, Georgia were used as a second geographical site. These ponds located on the piedmont plateau contained little, if any, submersed vegetation. Since the ponds contained established fish populations of the desired species, only enough fish were added to ensure an adequate number of fish for the experiment. Water hyacinths were transported from Florida for stocking in the ponds. Total coverage of the water hyacinths did not exceed 5% of the surface area of each pond.

TREATMENT

Ponds were sprayed with 2.24, 4.48, or 8.96 kg 2,4-D acid equivalent per hectare (2, 4, or, 8 lb/acre) using a commercial formulation of herbicide (Weedar 64®, Amchem Products, Ambler, Pennsylvania). All dilutions were made with water and no adjuvants were used. The Florida ponds were sprayed on July 12, 1971, and the Georgia ponds on July 26, 1971.

From Gangstad, E. O., Schultz, D. P., and Zeiger, C. F., *J. Aquatic Plant Management*, 14, 43, 1976. With permission.

TABLE 20.1

Residues of the Dimethylamine Salt of 2,4-D in Water, Hydrosoil, and Fish From Ponds in Florida and Georgia Treated with 2.24, 4.48, and 8.96 kg 2,4-D Acid Equivalent per Hectare

Pond	kg/ha	Depth (m)	Temp. (°C)	Time (days)	Water (mg/l)	Hydrosoil (mg/kg)	Fish (mg/kg)
Florida	2.24	1.3	34	1	0.025	0.005	0.080
	4.48	1.0	31	1	0.155	0.014	0.048
	8.96	1.2	31	1	0.312	0.033	0.005
Georgia	2.24	1.3	27	1	0.025	0.018	0.005
	4.48	0.9	29	1	0.233	0.024	0.014
	8.96	1.2	30	1	0.657	0.026	0.022
Florida	2.24	1.3	30	3	0.005	0.005	0.005
	4.48	1.0	30	3	0.172	0.014	0.005
	8.96	1.2	31	3	0.345	0.046	0.004
Georgia	2.24	1.3	29	3	0.087	0.008	0.005
	4.48	0.9	32	3	0.390	0.018	0.005
	8.96	1.0	30	3	0.692	0.040	0.005
Florida	2.24	1.3	31	7	0.005	0.005	0.005
	4.48	0.9	31	17	0.048	0.010	0.005
	8.96	1.2	31	17	0.025	0.008	0.005
Georgia	2.24	1.3	26	7	0.059	0.010	0.005
	4.48	0.9	27	7	0.400	0.018	0.005
	8.96	1.0	29	7	0.395	0.042	0.005
Florida	2.24	1.3	32	14	0.005	0.005	0.036
	4.48	1.0	32	14	0.005	0.010	0.005
	8.96	1.2	31	14	0.005	0.013	0.043
Georgia	2.24	1.3	28	14	0.027	0.005	0.005
	4.48	0.9	31	14	0.008	0.005	0.005
	8.96	1.0	30	14	0.050	0.005	0.005
Florida	2.24	1.3	30	28	0.005	0.005	0.005
	4.48	1.0	32	28	0.005	0.007	0.005
	8.96	1.2	30	28	0.005	0.005	0.005
Georgia	2.24	1.3	27	28	0.005	0.006	0.005
	4.48	0.9	31	28	0.005	0.005	0.005
	8.96	1.0	30	28	0.005	0.005	0.010

Summary of Treatment Means

					Water	Hydrosoil	Fish
Rates	2.24				0.020	0.006	0.018
	4.48				0.118[a]	0.010	0.009
	8.96				0.208[a]	0.018	0.009
Significant difference[a]					0.099	0.018	NS[b]
Days				1	0.235	0.020	0.029
				3	0.281	0.022	0.005
				7	0.155	0.016	0.005
				14	0.017[a]	0.007[a]	0.017
				28	0.005[a]	0.006[a]	0.006
Significant difference[a]					0.127	0.008	NS[b]

[a] P = 0.05.
[b] NS = not significant.

Data from a cooperative study of the U.S. Army Corps of Engineers and the Fish-Pesticide Research Laboratory, Bureau of Sport Fisheries and Wildlife, Warm Springs, Georgia, 1973.

SAMPLING PROCEDURES

Samples of water, hydrosoil, and fish were taken at 0, 1, 3, 7, 14, 28, 56, 84, 112, and 140 days after treatment. Fish were placed in live cages in the Florida ponds for the 1- and 3-day samples. Thereafter, fish were collected by hook and line, seine, or set line. They were wrapped in aluminum foil, bagged, and frozen on dry ice. Water samples were taken with a 2-l Kemmerer water bottle, and were composites of samples from shallow (0.3 m), medium depth (0.6 to 0.9 m), and deep (7 m or more) areas. Water for residue analysis was placed in quart jars, and acidified to a pH of less than 2 with concentrated sulfuric acid. The jars were capped with aluminum foil, and sealed with screw caps. Hydrosoil samples were taken with an Ekman dredge from shallow, medium, and deep sites. The three samples were composited for residue analysis. Samples were placed in plastic bags, frozen in dry ice, and kept frozen until analyzed.[22-24]

STATISTICAL ANALYSIS

The treatment effects are summarized in Table 20.1, using a pooled error of rates by dates as a test of significance.

WATER RESIDUES

The highest amount of 2,4-D residue in Florida pond water was 0.345 mg/l found 3 days after treatment in the pond treated at 8.96 kg/ha (8 lb/acre). Residue levels of 2,4-D found in Florida ponds decreased to 0.005 mg/l within 14 days after treatment. In Georgia pond water the highest detectable residue was 0.692 mg/l 3 days after treatment at 9.96 kg/ha. Only trace levels or no residues were detected 14 days after treatment.

HYDROSOIL RESIDUES

The highest 2,4-D residue detected in hydrosoil from Florida ponds was 0.046 mg/kg found 3 days after treatment at 8.96 kg/ha. In the Georgia ponds, the highest residue found was 0.042 mg/kg

on the seventh day after treatment at 8.96 kg/ha. The dissipation of 2,4-D from hydrosoil is due in great measure to microbiological degradation.

FISH RESIDUE

The highest residues found in any fish were samples from the 1-day harvest from ponds in Florida. These fish had been kept in live cages for up to 3 days after treatment to facilitate sampling. Hence, they were unable to escape from the applied herbicide. No residues were detected in fish from Florida ponds at the 3- or 7-day samples; however, negligible residues were detected in fish at the 14-day sampling. This may have been due to the release of the herbicide from decaying vegetation. Only one fish from the Florida ponds contained a detectable residue after 14 days and this was less than the negligible level.

The highest 2,4-D residue found in fish from the Georgia ponds was 0.075 mg/kg found in one of three bluegills harvested at 14 days from the pond treated at 8.96 kg/ha. No detectable residues were found in any fish from the Georgia ponds at the 3- or 7-day harvests, paralleling the results found in the Florida ponds.

None of the control fish contained detectable residues of 2,4-D, and the 2,4-D residues found in the exposed fish were well below the toxic levels for the dimethylamine salt formulations.

WATER HYACINTH CONTROL

The effect of the 2,4-D application on water hyacinth in the Florida and Georgia ponds was assessed by visual observation. Seven days after spraying, nearly all of the water hyacinth in all ponds was brown and decomposing. An estimated 98% of the plants were killed by the herbicide application, with no differences in kill noted among the different treatment levels. Since all three treatment levels of the herbicide were equally effective, it would be best to use the lowest effective concentration (2.24 kg acid equivalent per hectare) although retreatment may be necessary in some spots to prevent reinfestation.

Chapter 21

RESIDUE STUDIES IN SLOW-MOVING WATER*

LOUISIANA PROGRAM

The operations and maintenance project for the removal of water hyacinth and alligator weed in Louisiana has been in existence for many years, and the application of 2,4-D has evolved as the most effective means of control. The Corps of Engineers in 1966 began seeking to have 2,4-D dimethylamine registered for use on water hyacinth, including use in slowly moving water. In anticipation of the 1972 amendment to the federal Insecticide, Fungicide, and Rodenticide Act, information for this registration at an application rate of 2.24 to 4.48 kg/ha (acid equivalent per surface hectare) was submitted to the Registration Division, Environmental Protection Agency, Washington, D.C. on June 7, 1970. A tolerance was established and was published in the *Federal Register,* 40CFR180. subpart c(f), December 16, 1975, as 1 ppm in crops and raw agricultural commodities, fish, and shellfish.

Spraying policy with 2,4-D in potable water resources has been developed in accordance with instructions of the Louisiana State Department of Health, that no spraying would be done within 1 km of potable water intakes. A negligible residue in potable water has been established and published in the *Federal Register* as a food additive tolerance, Section 123.100, December 16, 1975. Although the aquatic plant control program is not intended for treatment of irrigation canals as such because Louisiana irrigation canals are privately owned and operated, benefits do accrue which are a result of water being free to move through main streams from which irrigation water is removed.

TIME COURSE STUDY

This study was conducted in a canal system owned and operated by the Southdown Corporation of Louisiana. This particular canal system was chosen because it provided a main canal which served as a common water source for the six lateral canals used an individual test locations. The main canal originated at Milton, Louisiana, and its water source was the Vermilion River. The plots con-

sisted of strips 166 (554.5 ft.) long by 3 (10 ft) wide, extending along opposite sides of the canal. The spray applications extended 0.61 m (2 ft) upon the canal bank and 2.44 m (8 ft) into the stream. This application method was selected to simulate an actual treatment situation where fringes of aquatic weeds are to be controlled along both sides of the stream. Water samples of approximately 0.75 l in volume were taken at each canal site, upstream, midplot, and downstream at 0.8, 1.6, and 3.2 km downstream from the point of application. Samples were analyzed according to standard methods of analysis by gas chromatography. The data are summarized for upstream, midstream, and downstream sites in Tables 21.1 and 21.2.

PUNTA GORDA PROGRAM

The city of Punta Gorda, Florida constructed a small water supply reservoir in 1964, impounding the waters of Shell Creek and Prairie Creek near that city. The reservoir had an approximate total surface area of 365 ha, 1 m average depth, and 3.4 million kl capacity. Conditions were favorable for growth of water hyacinth, and a total of 243 ha of the surface area was covered over with water hyacinth by September 1966. This created a serious problem of deteriorating water quality and fish and wildlife use. Numerous complaints were received by the Corps of Engineers, the Florida Game and Fresh Water Fish Commission, and the City of Punta Gorda. Several meetings were held with the Florida Department of Public Health to determine a suitable procedure for control. It was decided that spraying would not be undertaken within 1,000 ft of the intakes and that not more than 0.1 ppm residue would be permitted in the drinking water, a level that was acceptable for animal and human health.

RESIDUE ANALYSIS

Sampling stations were established within each arm of the reservoir and at the influent and effluent stations of the water treatment plant. It

Abridged and updated from Cooperative Research Studies of the University of Southwestern Louisiana, Lafayette and the U.S. Army Engineer District, New Orleans, 1975; and the U.S. Environmental Protection Agency and the U.S. Army Engineer District, Jacksonville, Florida, 1970.

TABLE 21.1
Treatment Conditions and Residue Data of 2,4-D Applied at 1.82 kg/ha for Control of Water Hyacinth in Rice Irrigation Canals in Louisiana

Site and canal[a]	volume dilution[b]	Temp. (°C)	Time (days)[c]	Residue (ppb)[d]
Upstream (1)	10	27	0.03	1
Upstream (1)	10	27	0.06	3
Upstream (1)	10	27	0.30	2
Upstream (1)	10	27	2.00	1
Midplot (1)	10	27	0.03	10
Midplot (1)	10	27	0.06	1
Midplot (1)	10	27	0.30	2
Midplot (1)	10	27	2.00	15
Downstream (1)	10	27	0.03	2
Downstream (1)	10	27	0.06	9
Downstream (1)	10	27	0.30	1
Downstream (1)	10	27	2.00	1
Upstream (2)	24	27	0.03	14
Upstream (2)	24	27	0.06	1
Upstream (2)	24	27	0.30	1
Upstream (2)	24	27	2.00	1
Midplot (2)	24	27	0.03	52
Midplot (2)	24	27	0.06	1
Midplot (2)	24	27	0.30	1
Midplot (2)	24	27	2.00	1
Downstream (2)	24	27	0.03	1
Downstream (2)	24	27	0.06	5
Downstream (2)	24	27	0.03	1
Downstream (2)	24	27	2.00	1
Upstream (3)	25	27	0.03	1
Upstream (3)	25	27	0.06	1
Upstream (3)	25	27	0.30	1
Upstream (3)	25	27	2.00	1
Midplot (3)	25	27	0.03	1
Midplot (3)	25	27	0.06	1
Midplot (3)	25	27	0.30	1
Midplot (3)	25	27	2.00	1
Downstream (3)	25	27	0.03	61
Downstream (3)	25	27	0.06	1
Downstream (3)	25	27	0.30	2
Downstream (3)	25	27	2.00	1
Summary of Treatment Means				
Upstream				2
Midplot				7[e]
Downstream				7[e]
Significant difference[e]				3
Days			0.03	16[f]
			0.06	2
			0.30	1
			2.20	2
Difference[f]				9

[a]Applied in a water solution at 1.87 kl/ha and 15 kl/ha and 15 kg/cm², sampled upstream, midplot, and 100 m downstream.
[b]Volume dilution of surface hectares applied per cubic meters per second of flow.
[c]Time in days after treatment.
[d]Parts per billion residue in the water 0.03, 0.06, 0.30, and 2.00 days after treatment, respectively.
[e]Significant difference: P = 0.05.
[f]Difference: P = 0.01.

Data taken from Time Course Study, a cooperative research study of the University of Southwestern Louisiana, Lafayette, Louisiana and the U.S. Army Corps of Engineers, New Orleans District, New Orleans, Louisiana, February 1975.

TABLE 21.2

Treatment Conditions and Residue Data of 2,4-D Applied at 3.63 kg/ha for Control of Water Hyacinth in Rice Irrigation Canals in Louisiana

Site and canal[a]	Volume dilution[b]	Temp. (°C)	Time (days)[c]	Residue (ppb)[d]
Upstream (4)	13	27	0.03	1
Upstream (4)	13	27	0.06	1
Upstream (4)	13	27	0.30	1
Upstream (4)	13	27	2.00	1
Midplot (4)	13	27	0.03	1
Midplot (4)	13	27	0.06	1
Midplot (4)	13	27	0.30	1
Midplot (4)	13	27	2.00	1
Downstream (4)	13	27	0.03	1
Downstream (4)	13	27	0.06	1
Downstream (4)	13	27	0.30	1
Downstream (4)	13	27	2.00	1
Upstream (5)	16	27	0.03	2
Upstream (5)	16	27	0.06	3
Upstream (5)	16	27	0.30	1
Upstream (5)	16	27	2.00	3
Midplot (5)	16	27	0.03	79
Midplot (5)	16	27	0.06	1
Midplot (5)	16	27	0.30	3
Midplot (5)	16	27	2.00	1
Downstream (5)	16	27	0.03	45
Downstream (5)	16	27	0.06	13
Downstream (5)	16	27	0.03	3
Downstream (5)	16	27	2.00	3
Upstream (6)	20	27	0.03	1
Upstream (6)	20	27	0.06	2
Upstream (6)	20	27	0.30	1
Upstream (6)	20	27	2.00	1
Midplot (6)	20	27	0.03	1
Midplot (6)	20	27	0.06	1
Midplot (6)	20	27	0.30	1
Midplot (6)	20	27	2.00	1
Downstream (6)	20	27	0.03	7
Downstream (6)	20	27	0.06	1
Downstream (6)	20	27	0.30	1
Downstream (6)	20	27	2.00	1

Summary of Treatment Means

Upstream				2
Midplot				8[e]
Downstream				7[e]
Significant difference[e]				3
Days			0.03	15[f]
			0.06	3
			0.30	1
			2.00	1
Significant difference[f]				9

[a]Applied in a water solution at 1.87 kl/ha and 14 kg/cm^2, sampled upstream, midplot, and 100 m downstream.

[b]Volume dilution of surface hectares applied per cubic meters per second of flow.

[c]Time in days after treatment.

[d]Parts per billion residue in the water 0.03, 0.06, 0.30, and 2.00 days after treatment, respectively.

[e]Significant difference: P = 0.05.

[f]Significant difference: P = 0.01.

Data taken from Time Course Study, a cooperative research study of the University of Southwestern Louisiana, Lafayette, Louisiana and the U.S. Army Corps of Engineers, New Orleans District, New Orleans, Louisiana, February 1975.

was determined that these sampling stations would represent approximately 1, 2, and 3 days after herbicide treatment. Water samples were taken in 3.7-l glass jugs according to standard methods, and chemical analyses of the water samples were made by the Environmental Protection Agency Region IV Laboratory at Athens, Georgia by gas chromatography. The results of treatments made in Shell Creek and Prairie Creek Reservoirs during April, June, July, August, September and December are summarized in Tables 21.3 and 21.4, respectively. At no time did the residue in potable water exceed the limit, and for practical purposes, was less than 10 ppb.

MATRIX ANALYSIS

The data of the Florida and Louisiana studies were assembled and analyzed for statistical significance by a stepwise multiple regression program[9] on a FORTRAN computer for six variables, i.e., rate of application, herbicide release after spraying, volume dilution after spraying, temperature in degrees Centigrade, time in days after treatment, and residue of 2,4-D in parts per billion. A summary of the analysis is given in Table 21.5. The multivariate regression is 0.9476, accounting for 89% of the variability. The estimated residue values are less than 1 ppb, 1 day after treatment, for the levels of herbicide applied in the Florida and Louisiana program, under general conditions of treatment. These values are substantially below the tolerance level of 100 ppb.

TABLE 21.3

Treatment Conditions and Residue Data of 2,4-D Applied at 2.24 kg/ha for Control of Water Hyacinth in Shell Creek Reservoir

Treatment data[a]	Volume dilution[b]	Temp. (°C)	Time (days)[c]	Residue (ppb)[d]
Apr 68	37	23	1	1
Apr 68	37	23	2	1
Apr 68	37	23	3	1
Jun 68	129	27	1	1
Jun 68	129	27	2	1
Jun 68	129	27	3	1
Sep 68	180	26	1	1
Sep 68	180	26	2	1
Sep 68	180	26	3	1
Dec 68	225	18	1	1
Dec 68	225	18	2	1
Dec 68	225	18	3	1
Apr 69	113	23	1	10
Apr 69	113	23	2	10
Apr 69	113	23	3	1
Aug 69	150	27	1	1
Aug 69	150	27	2	1
Aug 69	150	27	3	1
Sep 69	150	26	1	1
Sep 69	150	26	2	1
Sep 69	150	26	3	1
Dec 69	150	18	1	9
Dec 69	150	18	2	9
Dec 69	150	18	3	1

[a]Applied in a water solution at 1.87 kl/ha and 14 kg/cm².
[b]Volume dilution of surface hectares applied per total area of the reservoir at 1 m average depth.
[c]Time in days after treatment.
[d]Parts per billion residue in the water 1, 2, and 3 days after treatment.

TABLE 21.3 (continued)

Treatment Conditions and Residue Data of 2,4-D Applied at 2.24 kg/ha for Control of Water Hyacinth in Shell Creek Reservoir

Treatment data[a]	Volume dilution[b]	Temp. (°C)	Time (days)[c]	Residue (ppb)[d]
Apr 70	75	24	1	7
Apr 70	75	24	2	1
Apr 70	75	24	3	1
Jul 70	225	27	1	8
Jul 70	225	27	2	1
Jul 70	225	27	3	1
Sep 70	180	26	1	9
Sep 70	180	26	2	2
Sep 70	180	26	3	1
Dec 70	125	18	1	22
Dec 70	125	18	2	1
Dec 70	125	18	3	1

Summary of Treatment Means

April		3
June, July, August		1
September		2
December		4
Nonsignificant difference		NS
Days	1	6
	2	2
	3	1
Significant difference[e]		3

[e]P = 0.05.

Data from Punta Gorda Study, a cooperative research study of the Environmental Protection Agency, Atlanta, Georgia and the U.S. Army Corps of Engineers District, Jacksonville, Florida, February 1970.

TABLE 21.4

Treatment Conditions and Residue Data of 2,4-D Applied at 2.24 kg/ha for Control of Water Hyacinth in Prairie Creek Reservoir

Treatment data[a]	Volume dilution[b]	Temp. (°C)	Time (days)[c]	Residue (ppb)[d]
Apr 68	50	23	1	1
Apr 68	50	23	2	1
Apr 68	50	23	3	1
June 68	450	27	1	1
June 68	450	27	2	1
June 68	450	27	3	1
Sep 68	180	26	1	1
Sep 68	180	26	2	1
Sep 68	180	26	3	1
Dec 68	226	18	1	1
Dec 68	225	18	2	1
Dec 68	225	18	3	1
Apr 69	180	23	1	6
Apr 69	180	23	2	10
Apr 69	180	23	3	1
Aug 69	225	27	1	1
Aug 69	225	27	2	1
Aug 69	225	27	3	1
Sep 69	225	26	1	1
Sep 69	225	26	2	1
Sep 69	225	26	3	1
Dec 69	450	18	1	9
Dec 69	450	18	2	9
Dec 69	450	18	3	1
Apr 70	180	24	1	7
Apr 70	180	24	2	1
Apr 70	180	24	3	1
Jul 70	225	27	1	1
Jul 70	225	27	2	1
Jul 70	225	27	3	1
Sep 70	125	26	1	6
Sep 70	125	26	2	1
Sep 70	125	26	3	1
Dec 70	125	18	1	26
Dec 70	125	18	2	1
Dec 70	125	18	3	1

Summary of Treatment Means

April				2
June, July, August				1
September				1
December				3
Nonsignificant difference				NS
Days			1	3
			2	1
			3	1
Nonsignificant difference				NS

[a]Applied in a water solution at 1.87 kl/ha and 15 kg/cm^2, spring, fall, and winter.
[b]Volume dilution of surface hectares applied per total area of the reservoir, 1 m average depth.
[c]Time in days after treatment.
[d]Parts per billion residue in the water 1, 2, and 3 days after treatment.
Data from Punta Gorda Study, a cooperative research study of the Environmental Protection Agency, Atlanta, Georgia and the U.S. Army Corps of Engineers, Jacksonville, Florida, February 1970.

TABLE 21.5

Summary of Matrix Analysis

Correlation Matrix

Variable	Release	Dilution	Temp. (°C)	Time	Residue
Rate	0.0085	−0.2824	0.2013	−0.1764	0.0299
Release	1.0000	.0772	0.0137	0.1941	−0.9467
Dilution		1.0000	−0.4404	0.3822	−0.0936
Temp. (°C)			1.0000	−0.2828	−0.0211
Time				1.0000	−0.2098
Residue					1.0000

Analysis of Variance

	DF	Sum of squares	Mean square	F. ratio
Regression	5	13727.259	2745.452	242.666
Residue	138	1561.289	11.314	

Multavariate Equation[a]

Variable	Coefficient	Standard error
Residue	= 927.8879 (constant)	
Rate	+.0027	0.0043
Release	−0.9244	0.0274
Dilution	−0.0018	0.0031
Temp. (°C)	−0.0924	0.1055
Time	−0.3385	0.4148

Estimated Values

Rate (kg/ha)	Volume dilution	Temp. (°C)	Time (days)	Residue (ppb)
2.24	1.0	27	1	0.65
2.24	1.0	27	2	0.32
2.24	1.0	27	3	0.00
4.48	1.0	27	1	0.66
4.48	1.0	27	2	0.33
4.48	1.0	27	3	0.01

[a]Multiple regression = 0.9476; standard error of estimation = 3.3636.

Chapter 22

RESIDUES OF 2,4-D IN CROPS

INTRODUCTION

Herbicides, along with insecticides and other pesticides, are under attack by some scientists and many writers of feature articles in magazines and newspapers. As a result, considerable public apprehension and opposition has been aroused regarding the safety of using herbicides in or near water supplies. Apparently it is not generally known that most herbicides are nontoxic to mammals at the rates used and most are not hazardous to fish.

About 150 species of aquatic and semiaquatic marginal plants create weed problems in one or more aquatic situations in the United States and Canada as well as other parts of the world. The most recently available U.S. Census data reveal that there are more than 2 million ponds and small reservoirs, 189,000 mi of drainage ditches, and 173,000 mi of irrigation canals in the United States. Surveys have shown that all or most of these aquatic areas are infested by or susceptible to infestation by aquatic or bank weeds which interfere with the conservation, movement, or various uses of the water resources. According to the 1961 *Statistical Abstracts,* there are more than 175,000 km^2 of inland freshwater surface in the United States. These areas do not include the Great Lakes or any bodies of water less than 20 ha, or streams less than 0.5 km in width. Weeds are serious problems in shallow areas less than 3m deep in these large lakes and in sluggish stretches of the large streams.

HERBICIDE USE

Extensive use of herbicides has been necessary for many years to replace or supplement mechanical methods of controlling aquatic and bank weeds and preserve agricultural, navigation, and recreational uses of our water resources. In 1957, a careful survey by the Agricultural Research Service (ARS) and Bureau of Reclamation revealed that 3% of the 230 km of irrigation canals in 17 western states were infested with aquatic weeds. More than 75% of the 215,000 km of ditch banks were infested with bank weeds. In that year, 54% of the weed-infested canals and 80% of the weed-infested ditch banks were treated with herbicides at a total cost of about $8 million. Those weed control operations were evaluated by irrigation companies and resulted in estimated savings of labor, canal and structural damages, and crop losses valued at about $16 million. The net productive value of the water saved was estimated at $39,000,000, nearly five times the cost of weed control. Aquatic weed problems are especially critical in lakes, streams, and navigation channels in Gulf and Atlantic Coast states. A survey conducted in 1963 by the U.S. Army Corps of Engineers showed total infestations of 65,000 ha of water hyacinth, 40,000 ha of alligator weed and 85,000 ha of submersed weeds. The survey did not include farm ponds, canals, or drainage ditches. Since 1963, Eurasian water milfoil has invaded an additional 100,000 ha of water surface in North Carolina and Florida, and Florida elodea (*Hydrilla verticillata*) has spread over 50,000 ha of fresh water in Florida.

AQUATIC HERBICIDES

Aquatic herbicides, chiefly 2,4-D, have been used extensively for control of water hyacinth and certain other floating and emersed weeds in Florida and Louisiana since 1950. During the extensive, mostly unrestricted use of 2,4-D prior to 1967, no serious problems of injury to fish, livestock, or humans was apparent. During the years 1959 to 1962, about 100,000 acres of water hyacinth and alligator weed were sprayed with 2,4-D in the Corps of Engineers Aquatic Plant Control Program. That did not include the extensive herbicide usage by other federal, state, and local agencies and private individuals during those 4 years. The use of aquatic herbicides other than 2,4-D has been limited in southern states except Florida where diquat, endothall, Silvex,® copper sulfate, and dalapon are being used to a considerable and increasing extent. The highly successful aquatic and marginal weed control program of the Central and Southern Florida Flood Control District is an excellent example of what can be

By Frank, P. A., Plant Physiologist, U.S. Department of Agriculture, Denver, Colorado.

TABLE 22.1

2,4-D in Vegetable Crops after Irrigation with Water Treated with Dimethylamine Salt of 2,4-D

Treatment rate (ppm)	Residues (ppm)			
	Potatoes	Cucumbers	Peppers	Snap beans
Sprinkler (2 Days after Treatment)				
0.0	ND	0.01	–	–
0.02	ND	0.01	–	–
0.22	0.03	0.01	–	–
2.2	0.12	0.01	–	–
Sprinkler (Harvest)				
0.0	ND	0.02	ND	ND
0.02	ND	0.01	ND	ND
0.2	0.04	0.02	0.01	0.01
2.2	0.14	0.01	0.05	0.03
Furrow (7 Days after Treatment)				
0.0	ND	ND	–	–
0.2	ND	0.01	–	–
1.1	0.05	0.01	–	–
Furrow (Harvest)				
0.0	ND	ND	ND	ND
0.2	ND	0.01	0.01	ND
1.1	0.05	0.01	ND	0.01

Note: ND signifies "none detected."

Data taken from Cooperative Research Study No. 16 of the U.S. Department of Agriculture, Denver, Colorado, July 1970.

accomplished by extensive and careful use of available effective herbicides for control of aquatic and marginal weeds.

EVALUATION OF AQUATIC HERBICIDES

The ARS conducts a nationwide research program on the control of aquatic and bank weeds. About 80% of the research program is on evaluation of herbicides for weed control, residues in water, and effects on irrigated crops. In addition to the research by the ARS, less extensive research on aquatic herbicides is conducted or supported by other federal agencies including the Bureau of Reclamation, the Fish and Wildlife Service, and the Army Corps of Engineers. At least ten commercial companies screen chemicals for effectiveness of aquatic or bank weeds. State experiment stations or other state agencies in at least seven

states conduct evaluation programs on aquatic weeds.

STUDIES AT PROSSER, WASHINGTON

Six crops were planted in experimental plots a Prosser, Washington, and were sprinkler- and fur row-irrigated with water treated at several rates o the dimethylamine salt of 2,4-D. For the furro irrigation of crops, water containing the dimethy amine salt of 2,4-D at rates of 0.0, 0.02, and 2. ppm of 2,4-D was employed. The sprinkler irriga tions were made at rates of 0.0, 0.02, 0.22, an 2,2 ppm of 2,4-D, respectively. Analyses of appro; imately 100 water samples taken during the cour: of the irrigations confirmed the applied level Crop residues are summarized in Tables 22.1 an 22.2.

Cucumbers, sweet peppers, and snap bea

TABLE 22.2

2,4-D in Grain, Fodder, and Hay after Irrigation with Water Treated with Dimethylamine Salt of 2,4-D

Treatment rate (ppm)	Residues (ppm)				
	Sorghum		Alfalfa		
	Grain	Fodder	Green	Hay	Bean hay
Sprinkler (2 Days after Treatment)					
0.0	–	ND	ND	–	–
0.02	–	ND	ND	–	–
0.2	–	ND	ND	–	–
2.2	–	0.12	0.14	–	–
Furrow (Harvest)					
0.0	ND	–	–	ND	ND
0.2	ND	–	–	ND	ND
1.1	0.07	–	–	ND	ND
5.5	0.09	–	–	0.01	ND

Note: ND signifies "none detected."

Data taken from Cooperative Research Study No. 16 of the U.S. Department of Agriculture, Denver, Colorado, July 1970.

contained 0.05 ppm or less at all rates applied. Potatoes irrigated at the rate of 1.1 ppm, collected early and at fall harvest, contained 0.1 ppm or less of 2,4-D. Approximately 0.1 ppm of 2,4-D residue was found in sorghum grain and fodder and both green and dry alfalfa. Bean hay did not contain any 2,4-D. Analysis for bound 2,4-D, after extraction of free 2,4-D, indicated no measurable residues, and it is concluded that 2,4-D is present only as a negligible residue in the plant tissue.

Chapter 23
METABOLISM STUDIES OF 2,4-D IN FISH*

INTRODUCTION

The dimethylamine salt of (2,4-dichloro-phenoxy)acetic acid (2,4-D DMA) and the butoxy-ethanol ester of (2,4-dichlorophenoxy)acetic acid (2,4-D BEE) have been used extensively for the control of aquatic vegetation. The Tennessee Valley Authority used 888 tons of 20% 2,4-D BEE in 1966 on 8000 acres to control Eurasian water milfoil (*Myriophyllum spicatum* L.). Distribution and effects of the herbicide on these aquatic ecosystems were studied by Smith and Isom.[25] The application rates varied from 4 to 10 mg/l calculated on the basis of an average water depth of 3 ft. Mud samples contained concentrations ranging from 0.34 to 33.6 mg/kg of 2,4-D, whereas residues in bluegills (*Lepomis macrochirus*) were less than 0.15 μg/g. Studies by Wojtalik et al.[31] and Whitney et al.[30] indicated that when 2,4-D BEE was applied at the levels required for aquatic weed control, there were little or no effects on nontarget organisms of the ecosystem.

Mount and Stephen[17] found that the static, 96-hr TL_m (medium tolerance limit) for 2,4-D BEE for fathead minnows (*Pimephales promelas*) was 5.6 mg/l. Hughes and Davis[15] reported a 48-hr TL_m for bluegills of 2.1 mg/l. This ester was intermediate in toxicity to sunfish (96-hr LC_{50} of 3.6 ppm), i.e., it was comparatively less toxic than the butyl ester but somewhat more toxic than the iso-octyl ester and various salts and amines. Walker[28] also ranked their order of toxicity to various species of sunfish. Inglis and Davis[16] were unable to show significant effects of water quality on the toxicity of 2,4-D and other herbicides.

2,4-D BEE METABOLISM

Residue studies of the butoxyethanol ester of 2,4-D by Rodgers and Stalling[21] included the rate of hydrolysis of 2,4-D BEE in water, its uptake and elimination by fishes, and its distribution in organs of rainbow trout (*Salmo gairdneri*), channel catfish (*Ictalurus punctatus*), and bluegills. Extracts of tissues from bluegills and channel catfish exposed for 2 hr to ^{14}C-2,4-D BEE were examined

by thin-layer chromatography (TLC). Only the liver extracts contained sufficient 2,4-D BEE to be detected. The only other compound present was the 2,4-D acid. All other organs contained only the 2,4-D acid. These compounds were identified by TLC Rf values from autoradiograms. Harris hematoxylin and eosin preparations from 3-, 7-, and 14-day exposed fish did not differ from preparations of controls. Under the conditions of this study, no pathological effects were attributable to the exposure of fish to 2,4-D BEE or its metabolite.

Hydrolysis of 2,4-D BEE in water was rapidly accelerated by the presence of fish. A concentration of 1 mg/l of 2,4-D BEE was hydrolyzed in 24 hr when the aquarium tank was loaded with 1.5 g/l of fish. More than 90 hr was required for hydrolysis when fish were not present. Although the density of fish in lakes and ponds may be much less, plant growth is normally greater and the rate of hydrolysis may be increased by the activity of the biomass. In the field, 2,4-D residues in fish are often negligible 1 to 4 hr after exposure.[24]

Maximum whole body and tissue residues of 2,4-D are related to the concentration or availability of the ester, and they decline as the ester is hydrolyzed. Whole body residues were 7 to 55 times greater than the exposure concentration, 1 to 6 hr after treatment. The rapid uptake of 2,4-D BEE by fish may result from increased solubility and partitioning of the 2,4-D ester across the gill membrane. Absorption of the ester by lipids and lipoproteins in the blood may enhance the rapid distribution of the 2,4-D to various organs. In each species of fish tested, 2,4-D residues in tissues and organs rapidly attained a different maximum concentration, then the residues declined and exponentially approached negligible residues. However, if less than 1 to 3% of the ^{14}C-2,4-D were present as the ester, it would not have been detected in the organ extracts under the conditions of the experiment. Hydrolysis of the ester to the acid apparently takes place after the ester passes through the gill membrane, since detectable quantities of 2,4-D acid were not concentrated from the water.[21]

*By E. O. Gangstad and D. P. Schultz.

2,4-D DMA METABOLISM

The uptake, distribution, and dissipation of ^{14}C-labeled dimethylamine salt of (2,4-dichlorophenoxy)acetic acid (2,4-D DMA) from water by three species of fish were studied by Schultz[22] concurrently with the dissipation from water and hydrosoil. Fish were exposed to 0.5-, 1.0-, or 2.0-mg/l concentrations of herbicide for up to 84 days. Radioactive residues of the 2,4-D were determined by radiometric procedures in eight or more tissues and organs. Residues of 2,4-D were determined in muscle and whole body extracts by gas chromatography. Radioactive residues were found in all fish tissues and organs analyzed, but actual 2,4-D content was negligible in muscle, indicating that most of the ^{14}C residue was a metabolite(s) of 2,4-D. Residues of 2,4-D declined rapidly in water after 21 days, to less than 0.1 mg/l in 35 days, and in hydrosoil after 14 days, to less than 0.1 mg/kg in 21 days.

RADIOACTIVE RESIDUES

To learn why radioactive residues were accumulated in fish when the fish were exposed to ^{14}C-2,4-D DMA, the study by Schultz[22] was repeated. It was found that these residues did not reach maximum concentration, but continued to accumulate throughout a test period of 4 weeks. There was a significant loss of radioactive residue in extracts of fish exposed to ^{14}C-2,4-D DMA, if the sample was partitioned against an acid solution. This was also true for water extracts. These observations indicate that 2,4-D DMA was converted to carbon dioxide during the exposure period. The microflora associated with fish are assumed to be responsible for the degradation and recycling through fish flesh. To determine if fish exposed to ^{14}C-H$\overset{*}{C}$O$_3^-$ assimilated $\overset{*}{C}$O$_2$ into biochemical materials, rainbow trout were exposed to equal amounts of ^{14}C-NaH$\overset{*}{C}$O$_3$ and ^{14}C-2,4-D DMA for 2 weeks. The residue in fish from each experiment and the extent of 2,4-D degradation to H$\overset{*}{C}$O$_3$ during the 2,4-D DMA exposure was calculated (Table 23.1).

METHODS AND MATERIALS

Rainbow trout fingerlings (100 g total biomass) were exposed to 0.64 μg/ml of ring-labeled ^{14}C-2,4-D DMA in a 547-l fiberglass tank for 14

TABLE 23.1
Summary of 2-week Water Sample Radioactivity for ^{14}C-2,4-D$\overset{*}{,}$ $\overset{*}{C}$O$_2$, and ^{14}C-NaH$\overset{*}{C}$O$_3$[37]

| | ^{14}C-2,4-DMA Radioactivity (cpm/ml H$_2$O) | | |
Days	2,4-D and metabolites	H$\overset{*}{C}$O$_3$ radioactivity	^{14}C-NaH$\overset{*}{C}$O$_3$ radioactivity
0	534	0	750
7	422	–	300
14	326	105	230

Data from a cooperative study of the U.S. Army Corps of Engineers and the Fish-Pesticide Research Laboratory, Bureau of Sport Fisheries and Wildlife, Columbia, Missouri, 1973.

days. Water samples from the first day of exposure contained 534 counts per minute (cpm) per milliliter. However, 2 weeks later water samples contained only 326 cpm/ml (0.42 μg/ml). Several 14-day water samples were counted directly and then acidified, neutralized, and recounted. A 32% loss of activity was noted, suggesting that significant degradation of 2,4-D DMA to HCO$_3^-$ occurred. To determine the extent that HCO$_3^-$ might be involved in or responsible for ^{14}C residues in fish, an ancillary study was conducted using labeled ^{14}C-NaH$\overset{*}{C}$O$_3$ to which rainbow trout were exposed for 14 days. Treatment level and tank size were the same as the 2,4-D DMA study.

RESULTS AND DISCUSSION

Two week water samples showed a 59% decline in activity, indicating some exchange of CO$_2$ had occurred with atmospheric CO$_2$. Table 23.1 summarizes both water sample activity for 2,4-D DMA and NaHCO$_3$ exposures and the yield of $\overset{*}{C}$O$_2$ for the 2-week 2,4-D DMA exposure. Whole body samples of fish obtained from a 14-day exposure contained an average of 4130 cpm/g or 5.3 μg/g of 2,4-D and/or metabolites. Tissue samples extracted with diethyl ether gave an average recovery of 26% of the total tissue activity. The same tissue subsequently re-extracted with 1% phosphoric acid in methanol gave an additional 20% of activity recovered from the acid-methanol wash. However, over 50% of the total tissue activity was not recovered, and a loss of activity was observed when tissue samples were acidified.

Fractionation of the activity in the diethyl ether extract gave these results: 50 to 70% of the radioactivity was soluble in 1 N NaOH; of these base soluble materials, 60 to 70% were soluble in a dilute solution of NaHCO$_3$. Diethyl ether extract of whole body fish exposed to NaHCO$_3$ for week contained ca. 10% of the activity found in 2-week 2,4-D DMA exposed fish.

POND STUDIES

Experiments were designed to determine the residue levels and rate of dissipation of the dimethylamine salt of 2,4-D in water, hydrosoil, and fish from ponds treated at 2.24, 4.48, or 8.96 kg 2,4-D acid equivalent per hectare (2, 4, or 8 lb/acre).[22-24] These applications are 0.50, 1, and 2 times the recommended treatment rate. Ponds were located in two widely separated geographical locations to study the effects that different physical and chemical characteristics of the aquatic environment might have on the uptake and dissipation of the herbicide. The results of these studies of ponds in Florida and Georgia are summarized for the mean residues for rates of application and days after treatment in Table 23.2.

MULTIPLE REGRESSION ANALYSIS

For further study of the effects of different factors, the multiple correlation coefficients for the rate of application, pH, temperature, time in days after treatment, and the residues in waters and fish flesh were determined.[9] The results are summarized in Table 23.3. All factors were found to contribute to some degree to a multiple correlation coefficient, R = O.303, but temperature and the 2,4-D residue level in the water were the only factors of practical significance. These studies with 2,4-D BEE and 2,4-D DMA on rainbow trout, channel catfish, and bluegill demonstrate that residues of 2,4-D are not persistent and are not bioaccumulated in fish flesh by exposure to these herbicides at levels used in weed control treatments in ponds, lakes, reservoirs, marshes, bayous, drainage ditches, canals, or rivers and streams that are quiescent or slow-moving.

TABLE 23.2

Residues of the Dimethylamine Salt from 2,4-D in Water, Hydrosoil, and Fish from Ponds in Florida and Georgia Treated with 2.24, 4.48, and 8.96 kg 2,4-D Acid Equivalent per Hectare[38]

	Rate (kg/ha)	Time (days)	Water (mg/l)	Hydrosoil (mg/kg)	Fish (mg/kg)
Rates	2.24		0.020	0.006	0.018
	4.48		0.118[a]	0.010	0.009
	8.96		0.208[a]	0.018	0.009
Significant difference[a]			0.099	0.018	NS[b]
Days		1	0.235	0.020	0.029
		3	0.281	0.022	0.005
		7	0.155	0.016	0.005
		14	0.017[a]	0.007[a]	0.017
		28	0.005[a]	0.006[a]	0.006
Significant difference[a]			0.127	0.008	NS[b]

[a]Significant difference: P = 0.05.
[b]Not significant.

Data from a cooperative study of the U.S. Army Corps of Engineers and the Fish-Pesticide Research Laboratory, Bureau of Sport Fisheries and WIldlife, Warm Springs, Georgia, 1974.

TABLE 23.3

Summary of Regression Analysis

Correlation Matrix

Variable	Rate	pH	Temperature	Water	Days	Fish
Rate	1000	0.066	0.226	0.496	−0.114	0.156
pH		1.000	0.749	−0.032	−0.121	0.199
Temperature			1.000	−0.043	0.934	0.225
Water				1.000	−0.174	0.153
Days					1.000	−0.141
Fish						1.000

Analysis of Variance

Source	DF	Sum of squares	Mean square	F.
Regression	5	743.59	148.72	0.487
Residual	24	7328.27	305.34	

Variables in Equation

Variable	Coefficient	Standard error	F. to remove
Fish residue	(constant −49.651)		
Rate	0.020	0.137	0.021
pH	0.080	0.575	0.021
Temperature	1.977	1.327	0.366
Water	0.115	0.211	0.298
Days	−0.001	0.001	0.308

Summary Table

Variable entered	Multiple		Increase in RSQ
	R	RSQ	
Temperature	0.225	0.051	0.051
Water	0.278	0.077	0.026
Days	0.301	0.091	0.014
Rate	0.302	0.091	0.005
pH	0.303	0.092	0.008

Data from a cooperative study of the U.S. Army Corps of Engineers and the Fish-Pesticide Research Laboratory, Bureau of Sport Fisheries and Wildlife, Warm Springs, Georgia, 1974.

Chapter 24

RESEARCH ON INSECT ENEMIES*

STUDIES IN SOUTH AMERICA

Although the search for and preliminary testing of new enemies of water hyacinth is continuing in South America, present research is concentrated on five promising species of arthropods.

One of these arthropods is already present in the United States, apparently having entered this country along with water hyacinth. This species is an oribatid mite, *Orthogalumna terebrantis* Wallwork. It is known to be present in Uruguay, Paraguay, Argentina, Brazil, Surinam, Guyana, and Jamaica, and in the United States, in Florida and Louisiana. The mite has been found only on species of *Eichhornia* and *Pontederia*. The mites bore into and form 4 to 6-mm-long feeding galleries just below the upper surface of the leaves of water hyacinth.

Experiments are underway in Argentina (1971) to determine whether there are any significant differences between North American and South American populations of *O. terebrantis* which might warrant the introduction of the South American form into the United States.

A second arthropod, the bagoine weevil, *Neochetina bruchi* Hustache, is also being studied in Argentina. This species is recorded from Guyana, Brazil, Uruguay, and Argentina, feeding only upon species of Pontederiaceae. The adult weevils damage water hyacinth by surface feeding on the foliage. The weevil larvae tunnel and feed in the stem and crown of the plant, young larvae in the petioles, older larvae forming "feeding pockets" in the crowns. There may be from 1 to 12 larvae per plant. A rot usually follows the larval tunneling and the stem and leaf are completely killed. The larvae pupate underwater, forming cocoons from dead root hairs of the plant. *Neochetina* is apparently limited to a completely aquatic environment, although plants nearest the shore receive the most damage (see Figures 24.1 and 24.2).

In the laboratory starvation tests, *N. Bruchi* adults will feed upon several species of Commelinaceae, and on cabbage and lettuce. This feeding is not extensive and occurs only in the absence of pontederaceous plants; no feeding occurs on these plants if water hyacinth is present.[1]

Host specificity testing has been essentially completed and *N. bruchi* appears sufficiently specific to water hyacinth for introduction into the U.S. without danger to economic plants or native vegetation. Host specificity increased from adult feeding, to oviposition, to larval feeding. Adults fed and oviposited much less on other Pontederiaceae than on *E. crassipes,* and very slightly on the closely related Commelinaceae, lettuce, cabbage, and a few other plants in the aquatic habitat. However, larvae could complete their development only on water hyacinth except for one individual reared on *Reussia rotundifolia,* which does not occur in the U.S.[8]

Since it is proposed to introduce two very similar species to control water hyacinth, *N. bruchi* and *N. eichhorniae,* attempts were made to define their separate ecological niches. So far differences were found in seasonal abundance, seasonal rate of oviposition, preferred resting site of adults on the plant, preferred ovipositional site, manner of oviposition and location of eggs within the plant tissue, size and number of feeding spots, and small differences in host preference.

N. bruchi adults are much more abundant than *N. eichhorniae* throughout the spring and summer and early fall, but *N. eichhorniae* is most abundant in late fall and winter. *N. bruchi* reaches a peak population in March and *N. eichhorniae* in April and May. Peak larval populations occur in November and pupae in January. *N. bruchi* nearly always produces more eggs than *N. eichhorniae,* especially so in the spring; however, *N. eichhorniae* feeds more than *N. bruchi.* Both *N. bruchi* and *N. eichhorniae* feed and oviposit at all temperatures from 10 to 35°C, but maximum feeding and oviposition of both species occurs at 25°C. No parasites have been found so far on either species, but predators may possibly cause some mortality to eggs and larvae inside the stems.

The two species of *Neochetina* together cause considerable damage to water hyacinth in the field, with a maximum of 120 feeding spots per leaf, and 40% of the peticles damaged by larvae. In experiments in aquaria, *N. bruchi* killed all the

*By Coulson, J., Entomologist, U.S. Department of Agriculture, Beltsville, Maryland.

FIGURE 24.1. Dr. B. D. Perkins, USDA entomologist, showing a leaf attacked by the feeding of the water hyacinth weevil compared to normal leaves, in the laboratory at Fort Lauderdale, Florida.

water hyacinth plants quickest, *N. bruchi* and *N. eichhorniae* together next, and *N. eichhorniae* alone required the longest time.

Neochetina affinis, the only other species of the genus known in Argentina, appears about equally specific to the Pontederiaceae as *N. bruchi* and *N. eichhorniae.* Its natural hosts appear to be *E. azurea* and *Reussia rotundifolia,* and it is very rare on *E. crassipes.*

The water hyacinth mite, *Orthogalumna terebrantis,* was completely specific to water hyacinth in laboratory tests. Adults were found crawling over many of the test plants, but laid eggs only in water hyacinth. The immature stages develop within the living leaf tissue. Adult mites are almost unable to feed on living plant tissue, and probably feed on decaying tissue within the old galleries of the nymphs, in injuries, or in the feeding spots of

other insects and snails. In laboratory tests, adults survived longer on algae than on entire or injured leaves of water hyacinth. In the field, populations begin increasing in October, reach a peak in February, and practically disappear from water hyacinth in May. The adults spend the winter on *Pistia* where they probably feed on algae. Only a few larvae of *Acigona infusella* were found on water hyacinth. A population of *Epipagis albigutalis* was found in late October and again in February. A tiny dipterous larvae, *Thripticus* sp. was often numerous in the petioles of water hyacinth, but caused little damage.

A summary of the potential effects of insect enemies on the growth of water hyacinth is given in Table 24.1. It is postulated that these insect enemies will control water hyacinth in the United States.

FIGURE 24.2.   Close-up of water hyacinth weevil feeding on a water hyacinth leaf.

TABLE 24.1

Increment Factors of Water Hyacinth Increase in Various Locations with Hypothetical Number of Plants at this Rate after 90 Days, Starting with a Base of One Plant

Location	Daily increment factor to produce next count (= X of Bock's formula)	No. plants after 90 days (= N of Bock's formula, where T = 90 days	Time period of data	Ref.
California	1049	74	May 9–21, 1965	39
California	1056	134	May 21–June 7, 1965	39
Montevideo, Uruguay	1042	40	213 days, growing season, 1969 plants without insects)[a]	40
Buenos Aires, Argentina	1040	34	Summers, 1968[a] (without insects)[a]	41
Buenos Aires, Campana, and Santa Fe, Argentina	1000	1	(Natural enemies present)	41

Insects were removed from water hyacinth plants brought from the field to the laboratories. The plants were then liberated free of insects in laboratory ponds. During the period of the studies, distance acted as an adequate barrier against reinfestation by insects.

DOMESTIC STUDIES

Before introducing exotic natural enemies to control a weed, it is desirable to learn what natural enemies already attack that weed in the areas in which the introductions are to be made, what effect these species may have on the weed, and to what extent they may compete with or affect any introduced species.

To get information of this type, ARS supported a 4-year grant at Louisiana State University to study the insects associated with aquatic weeds of foreign origin in Louisiana. In 1969, an ARS entomologist surveyed water hyacinth in Florida, Louisiana, and eastern Texas, and recorded the natural enemies found. An entomologist from the Commonwealth Institute of Biological Control (CIBC) station in Trinidad has also recorded natural enemies of water hyacinth in Florida and Louisiana.

Of the 15 species of arthropods found during these studies to feed to some extent on water hyacinth in southeastern United States, only two are considered important enemies of the weed. One is the oribatid mite, *O. terebrantis,* discussed above. The other species is a noctuid moth, *Arzama densa* (Walker), whose larvae behave in much the same manner as the South American caterpillar *Acigona infusella,* that is, they tunnel and feed in the stems of water hyacinth.

Arzama densa is a native of North America, ranging from Maryland to Florida and along the Gulf Coast to Texas. Its original food plants, before the arrival of water hyacinth, were apparently the native North American species of *Pontederia.* Water hyacinth, being sufficiently closely related to *Pontederia,* has become a very satisfactory additional food plant for *A. densa.*

The biology and life history of *A. densa* and its importance as a control agent for water hyacinth have been studied extensively at Louisiana State University.[26,27] Vogel and Oliver of that institution report, and the ARS and CIBC investigations also indicate, that this species can be an important controlling agent for water hyacinth, but that, unfortunately, severe damage to the weed occurs only locally or sporadically. One factor limiting the effectiveness of *Arzama* is parasitism. Vogel and Oliver record five species of parasites of the egg, larvae, and pupae of *Arzama* in Louisiana, with a percentage of parasitism in the neighborhood of 50%.[19,20,26,27]

INSECT INTRODUCTION

The importation and interstate movement of plant-feeding organisms is regulated by the Agricultural Quarantine Inspection Division (AQI) of the USDA. Permits from AQI must be secured before any such movement can be made.

Although the final decision to allow or deny importation of an organism into the United States for the biological control of weeds rests with AQI, this division has for the past decade been aided in making this decision by a Subcommittee of the Joint Weed Committees of the U.S. Departments of Agriculture and of the Interior. The present composition of this subcommittee, called the Subcommittee on Biological Control of Weeds, includes representatives from the Forest Service and the Agricultural Quarantine Inspection, Plant Protection, Plant Science Research, and Entomology Research Divisions of the USDA, and the Bureaus of Sport Fisheries and Wildlife and Land Management of the USDI, and of the National Arboretum.

This subcommittee (1) reviews the adequacy of the specificity tests on organisms proposed for introduction, (2) suggests additional tests and studies if thought desirable to insure the safety of the introduction, (3) identifies and helps resolve conflicts of interest which sometimes arise in connection with the biological control of weeds, and (4) recommends for or against the proposed importation.

After an importation is approved by this subcommittee, permits are sought for the introduction. The Entomology Research Division also seeks approval from each state in which introductions of weed-feeding arthropods are planned. The concurrence of Canadian authorities is also solicited on a reciprocal basis.

Chapter 25

RESEARCH ON HOST SPECIFICITY

INTRODUCTION

One insect considered to be most promising from the standpoint of host specificity and potential for control of water hyacinth is the weevil *Neochetina bruchi* Hustache. During the course of study it was discovered that two species of weevils were involved, *N. bruchi* and a new species, *N. eichhorniae* Warner. A third species of *Neochetina* was found associated with *Eichhornia azurea, N. affinis* Hustache. The discovery of the two weevils delayed the program somewhat because some tests had to be repeated to be certain the right species was being used. Both species are apparently host specific to *E. crassipes.*[19,42]

A preliminary summation of the host specificity studies with *N. eichhorniae* was prepared and submitted to the Working Group on Biological Control of Weeds for clearance of *N. eichhorniae* into quarantine in the United States. The Working Group granted this request but recommended additional testing of several aquatic plant species before release in the United States could be considered.

TAXONOMY

Neochetina eichhorniae is a newly described species. The taxonomic breakdown is as follows:

> Class: Insecta
> Order: Coleoptera
> Family: Curculionidae
> Subfamily: Erirrhininae
> Tribe: Bagoini
> Genus: *Neochetina*
> Species: *eichhorniae*

Several members of the genus *Neochetina* are found feeding on members of the family Ponte-deriaceae, which includes water hyacinth. *Neochetina bruchi* Hustache is another species which is also found on *E. crassipes,* inflicting damage similar to that of *N. eichhorniae. Neochetina affinis* Hustache were collected on *Eichhornia azurea,* inflicting damage similar to that caused by *bruchi* and *eichhorniae* on *E. crassipes.* There appears to be a close evolutionary tie between the genus *Neochetina* and the family Pontederiaceae. The *Neochetina* spp. adults have a dull, grayish-brown appearance, looking not unlike a small lump of soil. The size is approximately 5-mm length and 2-mm width at the widest part. The various species of the genus have somewhat different markings and characteristics. *N. eichhorniae* has been recorded from Argentina, Bolivia, and Trinidad. Examinations of specimens from the field and from museum collections would likely indicate that *N. eichhorniae* is found throughout the range of *E. crasspies* in South America. The beetles used in these studies in Argentina were derived from populations in Santa Fe and Campana, Argentina.

LIFE CYCLE

Generally, the life cycle is as follows: The eggs are laid in October and November by the overwintering adults. The larvae (five instars) develop over a 3-month period during the summer months (November through January), and pupate in January. Adults emerge in January and February, producing eggs in March which give rise to a second generation. The larvae from these eggs feed and develop in the plants during March, April, May, and June. Pupae may be present in April and May, and adults may emerge in May and June. Thus, in May and June there may be an overlap of second-generation larvae and adults. The second-generation adults go through the winter. Therefore, two generations per year can be expected in the cooler regions, such as the vicinity of Buenos Aires; however, some of the second-generation larvae do not pupate, but continue into the winter. Quiescent larvae have been collected in the middle of winter within the crown of the plant. These have always been late instar larvae, usually fourth and fifth and never the first and second instars. It is probable that north of Argentina there is a complete overlap of generations due to continuous development.

Egg – The eggs are inserted individually or in groups in feeding scars of the adults, or in injured places on the petiole, usually on a protected area of the petiole, i.e., an area wrapped by a developing new leaf below the bulbous aerenchyma

By Perkins, B. D., Entomologist, U.S. Department of Agriculture, Fort Lauderdale, Florida, Technical Report 6, Appendix E, U.S. Army Engineers Waterways Experiment Station, Vicksburg, Mississippi, 1974.

tissue. As many as five eggs have been found in one oviposition spot. The eggs hatch in 7 to 10 days.

Larva — The typically white, apodous larvae immediately enter the petiole upon hatching. In several cases they have been found to exit from the original petiole, just above the crown, and then enter an adjacent petiole. One, or sometimes as many as five, late instar larvae can be found feeding within a pocket or gallery in the crown of the plant. A bacterial deterioration of the plant tissue surrounding the larval tunnel often accompanies the larval feeding, resulting in a sunken line on the petiole where the tunnel and surrounding aerenchyma tissue have collapsed. This often causes the petiole to die and separate from the plant, or become waterlogged, pulling the rest of the plant underwater where it dies.

Pupa — The larva moves to the outside of the plant and pupates in the underwater roots of the plant, forming a cocoon from cut pieces of the hyacinth root. This is a critical step in the development of the weevil, and the absence of hyacinth roots precludes completion of the life cycle. The pupal period lasts 20 to 30 days.

Adult — The emerging adults begin to feed almost immediately, causing small (2- to 4-mm diameter) feeding blotches on the leaf blades and petioles. The adults have a rather long life span with a maximum of 280 days recorded in laboratory rearing. The sex ratio is 1:1. Generally, the females are noticeably larger than the males within a given population.

During the day the adults are generally quiescent and can be found within the tightly folded tissues (petiole and young leaf shoots) above the crown of water hyacinth plants, a typical thigmotropic response. In the laboratory the insects have been found to wedge themselves into the crevices of tightly folded strips of moist cotton or gauze. They also crawl beneath pebbles and lumps of soil. There is also evidence of a thigmotropic interaction between the insects themselves. Often as many as eight weevils have been found huddled together in one spot on a plant. This is especially interesting since some plants in the same location, although fed upon, were barren of even a single insect within the tight tissue areas of the crown.

Adult feeding is usually nocturnal, although weevils have been found to feed on the petioles during the day. The nocturnal feeding and general diurnal quiescence may be due to the adverse effect of sunlight in drying and killing the adult insects. Studies have shown that they are very susceptible to heat, sunlight, and lack of moisture. Adult insects feeding on water hyacinth during the day have been found on plants with elongated petioles, but not on plants with the bulbous, floating petioles typical of uncrowded water hyacinth. This may be due to the tightly folded crown tissues being submerged underwater on the plants with elongated, less bulbous petioles.

Members of the weevil tribe, Bagoini, to which *Neochetina* belongs, are protected from the water by a characteristic wax coating over the body surface. The adults can withstand submergence for several minutes and on occasion were noted to walk down the stems and continue right on below the water's surface. Death from drowning is possible if they are submerged for prolonged periods. The insects are able to swim, although clumsily. Both the larvae and adults die quickly in the absence of adequate free moisture. A study in which weevils were caged with and without plant material under varying moisture conditions demonstrated this point.

NATURAL ENEMIES AND COMPETITORS

Among the enemies of *Neochetina eichhorniae* were several general predators. In the crowns of water hyacinth plants predaceous spiders, damselfly naiads, water tigers, and carabid beetles have been collected. These undoubtedly prey upon the soft-bodied larvae as they move from petiole to petiole. The adults, which are hard-bodied and larger, would be affected less.

The most important enemy so far is the fungus, *Beauveria* sp., which attacks the adults. It appears to attack the older, weaker overwintering adults and is found primarily in October and November. Laboratory populations affected with the fungus spot the cage with fecal material, indicating an upset digestive system and diarrhea, followed by a general weakened condition. The insect then becomes quiescent or paralyzed. In moist conditions the fungus quickly fills the body and becomes visible between the sclerites as a white, cotton growth.

Enemies of the pupal stage include two species of mites, as yet unidentified. One species is typical in form to the tick-like predaceous species, and the other is more similar in appearance to the hai follicle type of mite. The mites, however, are not considered an important factor in limiting the weevil population.

Plant quality is an important factor in the growth and development of *Neochetina*. In this regard, competitors can change plant quality to the point that the plants no longer provide suitable shelter for the adult weevils, although the leaves may still be acceptable as food. Among the competitors which produce this effect are two genera of scarab beetles, *Cyclocephala* and *Dyscinetus*. The damage produced by these species consists of a large central gallery which generally destroys a petiole and much of the crown. The resultant mass of plant material quickly becomes putrid. The plants infested with these insects have usually been devoid of *Neochetina,* whereas other plants in the same area at the same time, which have been free of attack by *Cyclocephala* or *Dyscinetus,* have had rather high populations of the weevils.

POPULATION STUDIES

There is considerable variation among the populations of *Neochetina* in Argentina. The variation depends on time of year and location. The adults, for example, have been as few as one insect for every ten plants during winter in Tigre, Argentina, or as numerous as three insects per plant in Santa Fe, Argentina, during the summer. During the summer in Tigre the population increased to more than one insect per plant. (During one examination the ratio in Tigre was 51 insects for 41 plants.) The adult populations will usually dwindle to almost zero by the end of October due to the effect of pathogens, principally *Beauveria,* on the aged, overwintering adults. In most localities all plants exhibit varying degrees of damage when the insects are present. Weevils of the tribe Bagoini, to which *Neochetina* belongs, are tied to an aquatic habit, and are characterized by a waxy coating which serves to waterproof them.

The genus *Neochetina* is closely tied to plants of the family Pontederiaceae, three known species of *Neochetina* being recorded only from the genus *Eichhornia* and *Pontederia*. Their distribution in South America is limited to that of *Eichhornia.* Both adults and larvae of *N. eichhorniae* live only briefly in the absence of free moisture. An aquatic or semiaquatic habitat is essential to their survival. Only *E. crassipes* serves as an adequate host for *N.*

eichhorniae. It can complete development only on this plant due to the specialized pupal cell, formed within the mass of floating roots. In the field, examination for larvae and/or adults of *N. eichhorniae* on aquatic plants growing in association with water hyacinth have indicated that the weevil only occurs naturally on *E. crassipes.*

SPECIFICITY TESTS

In addition to the observations above on biology and distribution, a series of feeding and behavior tests were conducted on a variety of plants. The plants designated for testing were selected on the basis of their taxonomic relation to water hyacinth, and the likelihood of their physical proximity to the weed in the United States. All plant tests were conducted in Argentina with the exception of *Nelumbo lutea* and *Nuphar advena,* which were tested at Buenos Aires.

Several types of tests were used:

1. Starvation test: in this test, the adult or larva caged with the test plant until the insect died. No choice of plants was offered.

2. Paired plant testing: test plants sustaining any nibbling or feeding damage in the starvation test were retested, each paired with a water hyacinth plant. This test indicated whether the weevil would damage other plants in the presence of its preferred host. Close confinement of the weevils with water hyacinth and the test plant in the paired plant tests limited the insects more severely than would normal field conditions. Thus, a third type of test, the "plant group test," was employed.

3. Plant group tests: plants nibbled notably in the starvation tests were further tested in the presence of water hyacinth under open and closed cage conditions. This test more closely simulated natural outdoor conditions in that the insects had a choice of plants and were free to leave the area when the cage was removed.

FEEDING TESTS

Larva

The larvae used in testing were unfed, first instar larvae obtained from eggs laid by adults originating in Santa Fe, Argentina. Thus, to obtain unfed first instar larvae, the adult insects were forced to oviposit on putrid water hyacinth

179

material. (In earlier tests it had been found that eggs were readily oviposited in moist, cut, and decomposing host material.) The neonate larvae, however, would not readily accept the putrifying material as food, and would crawl from the material. These larvae were carefully transferred to the test plants with a fine camel's hair brush, so as to reach the base of the plant. In no case did the larvae penetrate the test plants. In the first study the larvae even encountered difficulty penetrating water hyacinth.

A second study was set up in which the test plants were artificially punctured with the point of a pin and the larvae placed head first into the puncture. The plants accepted in the testing were all members of the Pontederiaceae, e.g., *E. crassipes*, *E. azurea*, and *Pontederia cordata*. One larva entered lettuce, but did not survive. The best survival was with *E. crassipes*, with one larva traveling 15 cm in each of two petioles.

Larval transfer tests with the plants *Sparganium*, *Nelumbo*, *Nuphar*, and *E. crassipes* were accomplished at Albany, California, using techniques similar to those described above. F_1 larvae from adult weevils received from Sante Fe, Argentina, were dissected from the leaves of water hyacinth and placed into cuts on the test plants. Out of ten larvae tested per plant species, eight were recovered alive on water hyacinth and none from the other plant species.

Adult

Starvation test — Preliminary trial tests had been conducted with plants in a soil substrate and plants with water exposed below the plant, but for uniformity, the enclosed upper plant was determined to be more desirable. The soil substrate proved inadequate since the insects could burrow into the soil and remain quiescent, but still alive, for prolonged periods of time. Plants caged over open water led to adults drowning when somewhat weakened by starvation. In both instances, results could be misleading (prolonged life with insects in the soil, or shortened life with insects drowned).

Adult feeding tests with the plants *Sparganium*, *Nelumbo*, *Nuphar*, and *E. crassipes* were accomplished with 10-mm-diameter leaf disks and/or 2.5-cm stem sections held on moist filter paper in petri dishes. After 9 days of feeding, all of the 30 adults on *E. crassipes* were alive. One egg was laid on *Nuphar* and four on *E. crassipes*.

Paired plant tests — Test plants on which some

adult feeding occurred in the starvation tests were caged together with *E. crassipes* and again exposed to beetles. These paired tests were done in 29.5- by 37- by 35-cm high aquaria serving as the cages. The same type of water-filled jars described in the starvation tests were used to hold the plants. However, the jars were capped with a sheet of plastic cut to fit exactly inside the aquarium in a horizontal position and which extended to the sides of the aquarium, where it was glued in place parallel to the aquarium bottom. This sheet of plastic served as a cap for the jar and also formed a false bottom halfway up the side of the test cage. The insects were introduced onto the center of the plastic floor with equal access to either the test plant or *E. crassipes*. Five insects were introduced per cage with the exception of one of the *Zebrina* replicates and four of the *Pontederia* replicates, each of which had ten insects. The only test plants evidencing feeding damage were *Zebrina*, *Pontederia*, and *Eichhornia azurea*. In all cases, the damage to the paired *E. crassipes* plant far exceeded the damage to the test species.

Plant group tests — To better approximate what might actually happen under field conditions, another test was devised in which all the test plants nibbled in the starvation tests were grouped in a large cage. Ten water hyacinth plants were placed in a small pool dug in the ground and lined with plastic. This pool was then covered by a 1-m-square screen-topped cage. Sixty adult *N. eichhorniae* were placed on the water hyacinth plants and the cage sealed all around the base with soil. Close around the outside of this cage were planted (opposite each other across the screen cage) two plants each of *Daucus carota*, *Lactuca sativa*, *Zebrina pendula*, *Impogandra elongata*, *Tradescantia crassifolia*, *E. azurea*, *Pontederia cordata*, *Commelina coelestis* and *Commelina virginica*. These were the test plants nibbled by *Neochetina* in the starvation tests noted above. The 60 adult weevils were allowed to become stabilized on the water hyacinth for 2 to 5 days. The following tests variations were then conducted consecutively.

Test 1 — The 1-m^2 cage was removed from over the water hyacinth and a second screen-topped cage 1.3 m^2 was then immediately placed over the entire test plot. Thus, the insects now had direct access to all of the test plants in addition to the water hyacinth. Observations were made daily except for weekends and field trip days. The plants were removed and examined 22 days after initiation of the test. The results of this test were as follows: Test plants damaged:

- *E. azurea,* 1 nibbled spot.
- All other test plants, undamaged.
- *E. crassipes,* 50 to 100 feeding spots per plant (10 plants × 50 to 100 = 500 to 1000 feeding spots).
- Number of adult *N. eichhorniae* recovered 38 out of 60 (30 living, 8 dead).

Test 2 – This test was set up the same as Test 1, except that after the time of stabilization of the 60 weevils on water hyacinth, both the inner and the outer cages were removed. Thus, the weevils were allowed access to all of the test plants and water hyacinth, or they could wander from the test area. The plants were examined 22 days after initiation of the test. Test plants damaged:

- All test plants, undamaged.
- *E. crassipes,* 200 feeding spots per plant (approximate) (10 plants × 200 = 2000 feeding spots).
- Number of adult *N. eichhorniae* recovered four living, one dead, six fragments.

Test 3 – This test was identical with Test 1, except after the removal of the 1-m² cage, the water hyacinth was treated with 2,4-D. The test plants, the insects, and the treated water hyacinth were then covered by the 1.3-m² cage. The test plants were examined after 14 days. The water hyacinth was brown and partially decomposed by this time. Test plants damaged:

- *Pontederia cordata,* 26 feeding spots.
- *E. azurea,* 25 feeding spots.
- All other test plants, undamaged.
- *E. crassipes,* 110, 240, 210, 120, 160, 170, 200, 140, 130 feeding spots per plant approximate) a total of 1580.
- Number of adult *N. eichhorniae* recovered, two on *P. cordata,* nine on *E. crassipes.*

Test 4 – This test was similar to Test 2 (i.e., both cages were removed), except that the water hyacinth was killed with 2,4-D. Two replicates of this test were made, each involving 60 adult weevils. It was believed a test of this nature would be important in assessing the danger potential to

plants growing alongside *Neochetina*-infected water hyacinths that might be treated with 2,4-D. Both replicates ran for 21 days. Replicate 1, test plants damaged:

- *P. cordata,* 19 feeding spots.
- *E. azurea,* 16 feeding spots.
- *E. crassipes,* over 100 feeding spots per plant (approximate) (10 plants × 100 = 1000 feeding spots).
- Number of adult *N. eichhorniae* recovered, five living.

Replicate 2, test plants damaged:

- *P. cordata,* two feeding spots.
- *E. azurea,* 9 feeding spots.
- *Zebrina pendula,* four tiny feeding spots.
- *E. crassipes,* 50 feeding spots per plant (approximate) (50 × 10 plants = 500 feeding spots).
- Number of adult *N. eichhorniae* recovered, eight living.

There is little evidence that plants other than water hyacinth will be attacked by the weevils as long as hyacinth is available. In the absence of suitable hyacinth, *E. azurea* and *P. cordata* may be attacked, although to a very limited extent. The feeding that occurred on these two plants in relation to that on water hyacinth can be summarized as follows:

	No. of feeding spots on	
	E. crassipes	*E. azurea* and *P. cordata*
Test 3	1580	51
Test 4		
Replicate 1	1000	35
Replicate 2	500	11

In all instances the beetles dispersed from the dying, sprayed plants, but failed to concentrate on nearby species on which they had fed under forced conditions. When the beetles are not confined, they will wander from even the healthy water hyacinth plants, as evidenced by the low recovery of adults in Test 2. This dispersal is probably the result of crowding, since under natural conditions, an average of six insects per plant is considered high.

Specimens of *P. cordata* were planted among *N. eichhorniae* infested hyacinth plants at the peak of the weevil population cycle. The *Pontederia* was attacked, but only to a very minor extent, the damage being negligible. This test was repeated when the weevil population was at a low ebb, with no evidence of attack on *Pontederia*. Of the other test plants placed in the field, only *E. azurea* was attacked, but again only to a relatively minor extent.

The potential effectiveness of *N. eichhorniae* as a control against *E. crassipes* was studied in the field and laboratory in Argentina. Weevil damage to water hyacinth can be grouped into three categories.

1. Larval feeding and tunneling
2. Adult feeding
3. Bacterial and fungal decay associated with feeding

The larval feeding effect is somewhat difficult to evaluate due to the covert location of the larva. Plant deterioration can be seen by a sinking of that portion of a petiole tunneled by the larva. A telltale browning of the tissue often follows this sunken appearance, beginning as a dark spot on the petiole at the entrance hole of the larva.

Frequently, the feeding of both the larvae and adults is accompanied by a fungal or bacterial deterioration of the tissue, which can cause a definite dieback of the hyacinth population. As many as one half of the petioles and leaves of a plant may be infected by fungus at a given time, the organisms apparently gaining entry through insect feeding scars. The affected tissues are brownish in color and eventually fall into the water/and become waterlogged. Definite dieback of the plant population occurs.

Chapter 26
CONTROL WITH PLANT PATHOGENS*

INTRODUCTION

Under a research contract with the University of Florida, and in cooperation with the Florida Office of Water Resources, all of the major lakes and rivers of Florida have been surveyed for diseases of water hyacinths (*Eichhornia crassipes*). Also included in these surveys were various locales in Georgia, Louisiana, Arkansas, Alabama, Tennessee, and the Chesapeake Bay area of Maryland. Overseas surveys have included Jamaica, Puerto Rico, Trinidad, El Salvador, Guatemala, Panama, Argentina, and India.[7,20,44]

Research thus far has resulted in the discovery of several diseases which could provide biological control of water hyacinth. The fungal blight from Panama, identified as being incited by *Rhizoctonia solani*, is a potential organism for this purpose. A fungus of water hyacinths in Florida, identified as *Fusarium roseum,* and a leaf-spotting disease of water hyacinths found in Florida and Louisiana, identified as being caused by a fungus in genus *Cephalosporium,* could also serve this purpose. A similar leaf-spotting disease was encountered at El Salvador, Jamaica, Trinidad, and Puerto Rico. Although this exotic pathogen induces zonate leaf spots on infected leaves, symptoms differ somewhat from material currently being screened for pathogenicity to water hyacinths. The most promising of all diseases for biological control of water hyacinths has been found in Rodman Reservoir, near Jacksonville, Florida. The significant symptoms are on the roots of infected plants.

During the past several years, it has been observed that serious disease conditions of water hyacinth are generally associated with various insect injuries to the plant. It is now believed that successful biological control will be best achieved by an insect-disease complex, and efforts are under way to study this possibility in greater detail.

FOREIGN EXPLORATION

During the last half of April 1973, a survey was made in Argentina and Uruguay in search of diseases of water hyacinth.[1] On this trip a rust disease of water hyacinth was found. The rust on

hyacinth was present in Argentina but not in Uruguay, where it had been reported to flourish. Morphologically, the rust fits the description of *Uredo eichhorniae.* This fungus was originally recorded as being found on water hyacinth in the Dominican Republic in 1927.[20] The organism has been loosely referred to as *Puccinia eichhorniae.* No basis for this name has been established, since only the uredial stage of this rust is known. The material from Argentina did not have teliospores, and they have not formed in material brought back to the plant pathology laboratory of the University of Florida at Gainesville.

Considerable work remains to be done to determine if this disease has biocontrol potential. It did not appear to be causing significant damage to water hyacinth in Argentina. However, this does not rule out the possibility that it will cause significant damage to the host plant in Florida. A different host gene pool may exist here that has not had the selection pressure of the disease exerted upon it. This we will soon determine. The big advantage of rust diseases is their host specificity. However, some rust species can affect related plant species and many have alternate hosts in nonrelated plant groups. Presently, we do not have information concerning additional related hosts or alternate hosts of *U. eichhorniae.* A similar rust also occurs on *Pontederia* in Florida. One additional exotic pathogen was added to the Florida collection. This was a species of *Alternaria* isolated from leaf spots on alligator weed from Puerto Rico (see Figures 26.1, 26.2, 26.3, 26.4 and 26.5).

ASSOCIATED FUNGI

The cataloging of fungi, both parasitic and saprophytic, that occurs in constant association with water hyacinth (Table 26.1) has been completed.

Many of the fungi found are of interest aside from their pathogenic capabilities in relation to the water hyacinth. For example, species of *Aspergillus,* especially *A. flavus,* and *Pithomyces* are known to produce mycotoxins in stored plant materials. The presence of these fungi may well

By Freeman, T. E., Plant Pathologist, University of Florida, Gainesville.

1. Lake Seminole
2. Lake Joseph, Lake Munson
3. Green Cove Springs
4. Lake Alice, Newnan's Lake
5. Lake Lochloosa, Lake Orange
6. Lake Oklawaha
7. Crescent Lake
8. Lake George
9. Crystal River
10. Lake Panasoffkee
11. Lake Dora, Lake Griffin
12. Lake Woodruff
13. Lake Monroe, Wekiva River
14. Lake Apopka, Lake Louisa
15. Lake Tohopakaliga
16. Hillsborough River
17. Lake Poinsett, Lake Washington
18. Lake Alfred, Lake Hancock
19. Lake Kissimmee
20. Crooked Lake, Lake Arbuckle
21. Lake Istokpoga, Lake Placid
22. Manatee River, Myakka River
23. Peace River, Prairie Creek
24. Lake Trafford
25. Loxahatchee National Wildlife Refuge

FIGURE 26.1. Collection sites in Florida for organisms pathogenetic to water hyacinth.

influence the potential utilization of water hyacinth as livestock food. Other saprophytic fungi are extremely important in biodegradation of water hyacinth mats that are under attack by other biological organisms or that have been treated with herbicides.

In 1974, the Rodman disease of water hyacinths was not as severe as in the two previous years. Survey trips begun in April were terminated in December with the advent of colder weather. The disease was evident in earlier trips, but as the year progressed there was a diminution in its prevalence. In previous years, it had been most damaging during the late summer and fall. However, by November of 1974, signs of the dysfunction were difficult to locate and by December none were discernible. All indications still point to a complex of organisms being responsible for the disorder.[7]

Initial results of field studies to assess the potential of domestic isolates of *Cephalosporium zonatum* and *Rhizoctonia solani* as biocontrols for water hyacinth were very encouraging. As noted in previous reports,[7,20] both fungi could be readily mass grown in liquid culture on a relatively simple chemically defined medium (Dzapek-Dox® broth amended with yeast extract). Both fungi were grown in this manner and the mycelial mats collected after 14 days. These mats were ground in a commercial size blender and the resulting suspension was sprayed onto water hyacinth plots in Lake Alice on October 10, 1975. The two fungi were applied singly and in combination. Infection by *R. solani* was apparent in less than 1 week and with *C. zonatum* in less than 2 weeks. By the first frost in early December, secondary spread of *C. zonatum* was apparent and resulting damage wa significant. Lesions caused by *R. solani* persisted

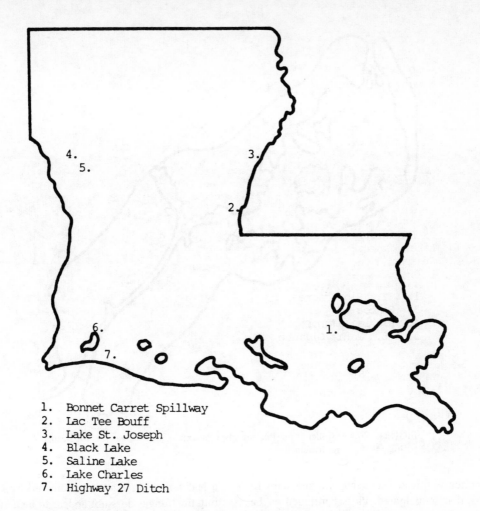

1. Bonnet Carret Spillway
2. Lac Tee Bouff
3. Lake St. Joseph
4. Black Lake
5. Saline Lake
6. Lake Charles
7. Highway 27 Ditch

FIGURE 26.2. Collection sites in Louisiana for organisms pathogenic to water hyacinth.

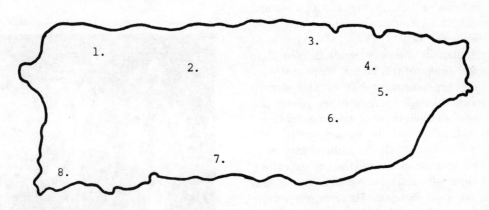

1. Lago de Guajateca
2. Lago Dos Bocas
3. Rio de la Plata
4. Aligibe Las Curias
5. Lago Loiza
6. Lago de Cidra
7. Lago Poncena
8. Laguna Cartagena

FIGURE 26.3. Collection sites in Puerto Rico for organisms pathogenic to water hyacinth.

FIGURE 26.4. Collection sites in the Panama Canal Zone for organisms pathogenic to water hyacinth.

1. & 2. Rio Chagres
3. Gamboa
4. Rio Bailamonos
5. & 6. Bahia Gigante

but secondary spread was negligible and damage was less than anticipated. Plots continued to be observed for renewed disease activity with the advent of warmer weather in the spring.

In addition to field studies with *C. zonatum*, various laboratory studies are also in progress. We are utilizing shortwave (253 nm) UV radiation to induce mutations that may result in even more pathogenic strains of this fungus. We presently are evaluating approximately 200 cultures derived from irradiated single spores for their growth and pathogenic capabilities. We have also obtained results indicating that the culture media upon which it is grown affects the pathogenicity of *C. zonatum*. Addition of yeast extract to the culture media apparently enhances pathogenicity of this fungus on water hyacinth. However, the culture medium alone also affects pathogenicity of the fungus. The fungus appears to be most pathogenic when grown on potato-dextrose broth amended with yeast extract (5 g/l).

An unidentified fungus isolated from water hyacinth in Manatee Springs proved to be pathogenic when reinoculated onto this plant. It causes a leaf lesion similar to that incited by *Fusarium*, but this fungus does not appear to be a member of this genus.

Cooperative studies with the USDA Laboratory

FIGURE 26.5. Diseased plants of water hyacinth, collected in January 1970, from the Chagres River, Republic of Panana. A, b, c, and d, are leaf-spotting zonate lesion of water hyacinth plants.

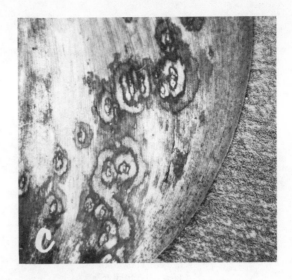

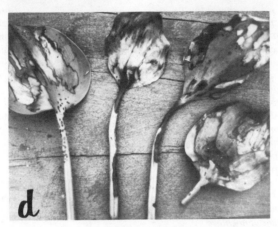

Figure 26.5 (continued)

in Fort Lauderdale, concerning the association of pathogens and the insects *Neochetina* and *Orthogalumna* have revealed that fungi belonging to several genera invade water hyacinths damaged by these insects. These secondary invaders amplify the damaging effects of the insects on plant populations.

FIELD STUDIES

During the second week in May of 1975, field studies using *C. zonatum* (culture 278) and *R. solani* (culture H287) were begun. Both fungi were grown in Roux bottles on liquid potato-dextrose broth amended with 5% yeast extract. At the end of 3 weeks with *C. zonatum* and 1 week with *R. solani*, fungal mats were collected, chopped in a blender, and each one was divided into two portions. One portion was sprayed onto crop plants growing in the field, while the other was sprayed onto water hyacinths growing in Lake Alice on the University of Florida campus. Both fungi infected water hyacinth, but infection by *R. solani* was disappointingly low. Infection by *C. zonatum* was relatively high and some indication of secondary spread was apparent. Crop plants were inoculated and the results are shown in Table 26.2. These results indicate that spraying water hyacinth with either of the fungi would probably not cause a serious threat to any of the crops tested.

INTEGRATED BIOCONTROL

In cooperation with the USDA Fort Lauderdale Laboratory, the effect of combined feeding by the weevil *Neochetina* and infection by *C. zonatum* is being studied. These two organisms have an additive effect that results in severe damage to water hyacinth.

TABLE 26.1

Fungi Found Associated with Water Hyacinth During Winter and Spring Months

Fungi	Pathogenicity
Subdivision	
Ascomycotina	
Class – Pyrenomycetes	
1. *Melanospora* sp.	None
Class – Loculoascomycetes	
2. *Leptosphaerulina* sp.	None
3. a. *Didymella exigua*	None
b. *Ascochyta* (imperfect stage)	
4. *Mycosphaerella* sp.	Slight
Deuteuromycotina	
Class – Coelomycetes	
5. *Phoma* spp.	Moderate to none
6. *Botryodiploidea* sp.	Slight
Class – Hyphomycetes	
7. Unknown synnematal fungus	None
8. *Mycoleptodiscus terrestris*	Slight to none
9. *Myrothecium cintum*	None
10. *Epicoccum purpurascens*	None
11. *Alternaria* spp.	None
12. a. *Aspergillus flavus*	None
b. *Aspergillus niger*	
13. *Acremonium* (Cephalosporium) *zonatum*	Good – excellent
14. *Bipolaris* spp.	Under investigation
15. *Cercospora* sp.	Good
16. *Cladosporium* sp..	None
17. a. *Curvularia brachyspora*	None
b. *Curvularia penniseti*	Slight
18. *Dendryphiella infuscans*	Not tested
19. *Exserohilum prolatum*	Under investigation
20. *Memnoniella subsimplex*	Not tested
21. *Perionia echinochloae*	Slight
22. *Pithomyces chartarum*	None
23. a. *Nigrospora oryzae*	None
b. *Nigrospora sphaerica*	None
24. *Thysanophora longispora*	None
25. *Scolecobasidium humicola*	None
26. *Stemphylium vericarium*	Under investigation
27. *Ustilaginoidea* sp.	Not tested

Data from Biological Control of Water Weeds with Plant Pathogens, a cooperative research study of the University of Florida and the office of the Chief of Engineers, Washington, D.C., 1973.

TABLE 26.2

Infection of Crop Plants by *Cephalosporium zonatum* and *Rhizoctonia solani* after Spray Inoculation with Mycelial and Spore Suspensions

Crop	Variety	Infection	
		R. solani	C. zonatum
Cabbage	Charleston Wakefield	–	–
Canteloupe	Hale's Best Jumbo	++	–
Carrot	Imperator	–	–
Celery	Giant Pascal	–	–
Collard	Georgia	–	–
Cucumber	Poinsett	+++	–
Eggplant	Florida Market	–	–
Endive	Green Curled	–	–
Escarole	Batavian	–	–
Field corn	PAG 751	–	–
Grapefruit	Duncan	–	–
Irish potato	Sebago	–	–
Lettuce	Great Lake	–	–
Lima bean	Henderson	–	–
Mustard	Florida Broadleaf	–	–
Oats	Fulghum	–	–
Okra	Clemson Spineless	–	–
Onion	White Globe	–	–
Orange	Temple	–	–
Peanut	Florunner	–	–
Pole bean	Kentucky Wonder	+	–
Radish	Scarlet Globe	–	–
Rye	Weser	–	–
Slash pine	Native	–	–
Snap beans	Harvester	–	–
Southern peas	Cream 40	–	–
Soybeans	Bregg	–	–
Squash	Early Summer Crookneck	–	–
Strawberry	Florida 90	–	–
Sugarcane	CL 41-223	–	–
Sweet corn	Silver Queen	–	–
Sweet pepper	Yolo L	–	–
Tangerine	Dancy	–	–
Tobacco	Turkish NN	–	–
Tomato	Homestead	–	–
Watermelon	Congo	+	–
	Charleston Grey	+	–
Wheat	Holden	–	–

Note: –, no infection; +, mild infection; ++, moderate infection; ++++, severe infection.

Data taken from Biological Control of Water Weeds with Plant Pathogens, a cooperative research study of the University of Florida and the Office of the Chief of Engineers, Washington, D.C., 1973.

Chapter 27

ENVIRONMENTAL IMPACTS

Most of the major pest problems in this country, as well as in other countries, are the result of accidental introductions of plants, insects, or animals, or of well-meaning purposeful introductions that backfired. There are several well-known examples of the latter: prickly pear cacti and rabbits into Australia (the first to provide a cochineal dye industry for the newly settled British colony, and for hedgerows, cattle food for drought seasons, and simply as an ornamental plant, and the second to provide sport for hunters); the mongoose into the West Indies (for snake control); and the English sparrow and starling into the United States (by an English bird lover). A recent news article in Florida mentions problems with tropical fish, introduced by fish fanciers which have escaped (or been dumped) into Florida streams and rivers. Other pest introductions resulting from the tropical fish industry should be added to this list — exotic aquatic plants, some of which upon escape have become most important aquatic weeds. Water hyacinth, though not an aquarium plant, is in itself another example of such an introduction, having been originally introduced as an ornamental, because of its beautiful blossom.[8,43]

BIOLOGICAL INTRODUCTION

These accidental or purposeful introductions, for the most part, were made prior to federal laws passed in the early part of this century designed to limit or control such introductions, and certainly prior to the current widespread awareness of environmental or ecological implications of such introductions and the resulting safeguards. The introduced plants and animals mentioned above as examples of "bad" purposeful introductions have all become pests in their new homes because their populations could not be readily controlled. In all cases, they were introduced without the biological agents responsible for regulating their populations in their homelands, i.e., without their natural enemies. These examples, therefore, are the direct opposite of introductions for biological control. Biological control entails, of course, the introduction of natural enemies, specifically to assist in controlling "bad introductions gone wild."

The first use of natural enemies to control pests in the United States concerned insect pests and occurred in the latter part of the 19th century. The first attempt to control weeds with natural enemies took place at the beginning of this century in Hawaii. Lantana, a tropical (Central American) plant, had been widely spread throughout the world as an ornamental. In Hawaii, and several other countries where it had been introduced, lantana escaped cultivation and rapidly became a serious pest weed. Introductions of insects into Hawaii from Central and South America began about 1902, but the main thrust of the program occurred about 50 years later. As a result of these introductions, lantana is now considered to be substantially controlled in Hawaii; results in other areas of the world have not been as uniformly successful.

Likewise, the prickly pear cactus problem in Australia was solved by the introduction of insects from South America. These introductions were begun in 1920, and by 1940, 95% of the land infested with prickly pears was reclaimed, the cactus being reduced to an incidental pest status. (Incidentally, the large populations of rabbits, which had been introduced into Australia, were also controlled by the introduction of a biological control agent, a disease specific to rabbits, myxomatosis.)

St.-John's-wort, or Klamath weed, an introduced noxious rangeland weed in the western United States, has been completely controlled in California, and substantially controlled elsewhere, by the introduction of insects which began in 1944. Within a decade, Klamath weed was reduced to 1% of its former abundance in California and is no longer a problem there. Savings over previous losses caused by this weed and prior costs of its control have been estimated to be somewhat over $2,000,000 a year.

Biological control of aquatic weeds is a relatively new endeavor, the first attempt being made against alligator weed, with which the Corps of Engineers, which has financed the program, is completely familiar. This work began in 1959, with the first release of insects beginning in 1964. Some success is now apparent in controlling alligator weed in some areas of the Southeast as a

By Coulson, J., Entomologist, U.S. Department of Agriculture, Beltsville, Maryland, Technical Report 10, Appendix F, U.S. Army Engineers Waterways Experiment Station, Vicksburg, Mississippi, 1974.

result of this program, especially in northern Florida, and most recently in Louisiana.

The most important points concerning these examples of successful biological control of weeds are (1) that if completely successful, the introduced control agents exert a continual control — no additional expenditures and no annual control efforts are required, once the control agents are well established and an integral part of the ecosystem; and (2) in no cases have the target weeds been eradicated, nor was eradication expected. Their populations have simply been reduced to a point where they no longer present an economic threat. An equilibrium between the target plant and its natural enemies is reached, *if* biological control is completely successful. Biological control is also considered partially successful even if the introduced natural enemies and introduced weed do not reach such an equilibrium; any reduction in the weed problem by natural enemies is an aid in weed control. A third point is that in the cases mentioned above the introduced control agents have not become pests in themselves, a fact that is the result of much detailed study made prior to their initial release.

CONFLICTS OF INTEREST

In considering biological control, the first question to be answered is, "What is a weed?" It is not as simple as it may sound. To an individual farmer or herdsman there is no question as to the "weediness" of certain plants in his own fields, pastures, or ranges that are detrimental to his interests, and that are causing him economic losses by reducing, damaging, or otherwise limiting the value of products produced on his lands. It is to his economic advantage to eliminate those plants causing those losses, by digging them up, plowing them under, burning them, or spraying them. His actions generally affect only his own property (although certain of these actions *may* have long-range, and sometimes immediate, effects, over a wider area; this is especially true of weed control programs that cover wide areas).

In the case of the use of introduced insects in weed control, a different situation is encountered. Insects and other biotic natural enemies are mobile; they cannot readily be confined to a small area, but, if successfully established, can spread to any area where food is available. In the case of insects introduced for biological control of weeds this food is a specific plant, the target weed.

Indeed, it is not the intent of biological control to confine the effects of control over a small area.

Therefore, what are termed "conflicts of interests" sometimes occur in the field of biological control of weeds. A plant that is classed as a weed in one area, or by some interests within an area, may be considered a beneficial plant in another area or by other interests within the same area.

One example will serve to illustrate such "conflicts of interest." Yellowstar thistle is an introduced plant that causes economic losses in grazing ranges and grain and seed crops in California. At the same time, this thistle is reported to be a key plant in the maintenance of the California bee industry, and thus beneficial in assuring adequate pollination of California fruit and seed crops. Very often fish and wildlife or horticultural considerations are also involved in these conflicts of interest.

Biological control can proceed only when any such conflicts of interest are resolved. In some cases, these are never resolved; in other cases, compromises are reached. In the case of yellowstar thistle, the cattle industry interest was judged to be the predominant direct interest, and biological control efforts are now under way. Reginal Plant Boards and special Federal-State Interagency Committees can be of considerable help in resolving such conflicts.

In the case of water hyacinth, there has never been any question of any conflict of interest.

Once any conflicts of interest are resolved, a biological control program can be initiated. Searches are made for the natural enemies responsible for regulating the weed's population in its native home. Insects found to cause significant damage to the weed are then subjected to rigorous tests to determine their feeding habits and specificity, and investigations are conducted to learn other details of their biology.

Much has been written on the subject of host specificity of plant-feeding insects, and it is discussed briefly here. It is known that insects can have either a wide range of host plants (polyphagous), a narrow range of closely related host plants (oligophagous), or be restricted to a single host plant (monophagous). By host plant is meant the plant species required in order for the insect to successfully complete its development and reproduce. The goal of biological control studies is, of course, to find a monophagous natural enemy that produces significant damage to the target weed.

After research has been completed providing

the experimental evidence required to assure the specificity and probable utility of a natural enemy for use in a biological control program, procedures are set in motion to release, establish, and evaluate the natural enemy as a biological control agent in a new home. These procedures include seeking international, national, and state review, and recommendations or approvals for the introduction.

ENVIRONMENTAL EFFECTS

Therefore, in the cases of introduction of foreign insects for the control of exotic weeds, many years of study are spent, and many knowledgeable experts in various disciplines and with various interests to protect are consulted, to assure that any effect of a proposed introduction on the environment will be that of *correcting* an existing detrimental effect resulting from earlier accidental or ill-advised introductions. These earlier introductions have without exception not been the subject of the rigorous research and clearance procedures characteristic of biological control projects.

It is the objective of biological control to restore, at least to some degree, the balanced conditions that existed prior to the introduction of the target pest. The introduced pest, by virtue of the fact that is was introduced without its own complex of natural enemies which served to limit its abundance in its native home, has a natural competitive advantage over the native plants, with their complex of natural enemies, in the environment in which the pest is introduced. A complete elimination of the pest is, of course, not the objective of biological control. But by decreasing the competitive advantage of the introduced pest through addition of the stresses caused by its own natural enemies, it is expected that a more balanced ecosystem will result in which the introduced weed will play only a small, nonpest role, or that, at least, the need for other control measures will decrease.

After 70 years of biological control weed research, no mistakes have yet been made, anywhere in the world, even in areas with less rigid importation regulations than the United States. There have been many introductions that have fizzled, that is, that have either failed to become established or have failed to have any serious effect on the abundance of the weed for which the introductions were made. But no insect introduced for weed control has ever become a pest. This excepts minor temporary feeding on crop plants which has occurred in three or four cases in the past, generally as a result of starvation feeding of populations of the introduced insect in the temporary absence of sufficient host plants.

Although it is experimentally possible to accurately predict the feeding range of insects proposed as introductions for weed control, it is impossible to accurately predict the degree of success of the introductions, that is, their final effect on the environment.

A prediction of the outcome, and the success of an introduction, depends on many factors.

1. It depends on the successful establishment of the introduction, which is by no means always assured.

2. It depends on how well the introduced natural enemy adapts to its new environment, and, often most important, its biotic potential in response to that new environment. Climatic and other limiting factors, including native parasites and predators, are often imposed on the introduced control agent that did not exist in its native home. Conversely, some of the limiting factors, especially parasites and predators in the native home of the weed, may not be as influential in the new environment. This is, of course, the hope of a biological control program. Insects to be introduced are carefully screened to eliminate any of their own natural enemies, i.e., predators, internal parasites, or pathogens. Abiotic limiting factors, whether added or substracted, are more difficult and often impossible to assess.

3. Success of an introduction depends upon its effect on the target pest. Much of this is learned during prior studies, both in the field in the native home of the weed and its enemies, and in later laboratory studies. But the effects on the target pest in the new environment in turn depend on the factors already mentioned, i.e., adaptability, biotic potential, native parasites and predators, etc., and sometimes on changes that have occurred in growth habits of the target weed itself in response to new environmental conditions.

Different combinations of these various interacting factors result in different degrees of success in different areas. In all cases of successful biological control of weeds, success has not been an all-encompassing or uniform success. Certain ecological situations occur in which the introduced control agent does not exhibit the same effect (or may not even become successfully established) as in other areas, where it may completely control the weed. In all cases, a complex of natural enemies has been required, some exhibiting control in one ecological situation, or geographical area, others becoming important in other situations or areas. Often a single insect can provide control over quite a large area, for example, the Klamath weed beetle in California, yet other insects, with other ecological requirements, are needed for control in other areas, such as for Klamath weed in British Columbia. The introduced alligator weed flea beetle provides another example, being limited in its effect not only by climatic limitations related to its own life cycle, but apparently also by certain nutritional deficiencies of its host plant growing under marginal conditions in pockets within the geographical area in which the beetle should be expected to provide some control.

POTENTIAL FOR CONTROL

In considering the potential effects of the *Neochetina* release, it is first necessary to consider the effects on the environment of water hyacinth itself. The economic losses caused by this plant, the number one aquatic weed pest in the United States, are well known, especially by the Corps of Engineers, which is charged with the task of controlling it in the waterways of the southeastern states. Other federal and state agencies in the Southeast charged with maintaining recreational and fishing waters clear of noxious weeds are also well aware of the problems caused by water hyacinth and the cost of controlling it.

It is a comparatively easy task to indicate what the effect on the environment would be if water hyacinth were eliminated or adequately controlled. And it is also simple to point out the advantages of biological control over currently practical chemical and mechanical control measures.

Should the introduction of *Neochetina eichhorniae,* with or without additional introductions of other South American natural enemies, successfully provide the stresses needed to significantly reduce the aggressiveness of water hyacinth in the United States and produce an ecological balance similar to that existing in South America, the beneficial environmental effects will include: an opening of large areas of water to recreational uses, an increase in the efficiency of water distribution and flood control systems, a reduction in areas of high mosquito breeding potential, an increase in the available oxygen supply of the water, and an increase in the utility of water for potable irrigation, and fish and wildlife uses, while at the same time, a reduction in or elimination of the weed for chemical and/or mechanical methods of control now practiced.

Concerning adverse environmental effects, it is again relatively easy to predict any caused by the potential elimination of large populations of the target weed. For water hyacinth, these are few. The only redeeming feature of the plant seems to be its beautiful blossom, a large number of which would be lost, and thus an aesthetic loss could perhaps be claimed, if the plant were eradicated. There are no claims that water hyacinth is of any great value to wildlife, to fish (to which it is definitely detrimental), or to livestock, or has any other value to man other than perhaps as an occasional ornamental for artificial ponds (a use which created the water hyacinth problem in this country in the first place).

Under potential adverse effects must be placed those possible short-term effects of the introduced natural enemy itself, which again depends on its successful establishment and its degree of success in providing control of the target weed. In the case of *Neochetina,* for example, possible short-term effects include a temporary increase in the amount of organic matter in waterways, if the weevil becomes responsible for the death of a large number of water hyacinth plants. There also exists the possibility that *Neochetina* might temporarily feed to a limited extent on pickerelweed (a food plant for fish) in some areas, if a population increased in excess of that needed to control the water hyacinth. *Neochetina* was never found to feed on other than species of *eichhornia* in nature in Argentina. At the same time, it can be pointed out that any possible damage by the weevil to stands of pickerelweed in association with water hyacinth in the United States will not be as extensive as damage to pickerelweed stands already being caused by current water hyacinth control practices.

Alternatives to the introduction of natural enemies for water hyacinth control consist of the following.

1. Continuation of chemical or mechanical controls, or a combination of the two, either as currently practiced or as altered by results of on-going or future research in these fields.

2. Possible mechanical harvesting and utilization of water hyacinth, the feasibility of which is yet to be demonstrated.

3. Possible utilization of native natural enemies. There are two arthropod enemies of water hyacinth that already occur in the United States, both of which also attack pickerelweed. These are the moth *Arzama densa* Walker and the mite *Orthogalumna terebrantis* Wallwork. Neither are at present particularly effective in providing any substantial control of water hyacinth, but the possibility exists that they might be utilized in mass release programs. Long-term research would be required to develop means for utilizing these arthropods in such a mass control program, and the effectiveness of such a program is subject to doubt. Another native enemy of water hyacinth is the manatee, a general feeder on a wide range of aquatic plants which exists only in Florida. Various studies have already indicated that severe problems exist in the possible utilization of this mammal for aquatic weed control, including dif-

ficulties in its propagation, in protecting it, and in the fact of its localized distribution.

4. Possible utilization of introduced fish or snails in the control of a wide range of aquatic weeds, including water hyacinth; although research is underway, full evaluation has not been completed on the techniques of rearing and utilization of these control agents or on the environmental effects inherent in their use.

5. The final alternative is simply the application of no control at all, an alternative that is universely considered unacceptable in the case of water hyacinth.

These, then, are the potential effects of (and alternatives to) the introduction of foreign natural enemies for water hyacinth control in the United States, of which *Neochetina eichhorniae* is but the first of several proposed for introduction. Eleven years of research and nearly $400,000 have been expended, and the opinions and recommendations of scientists of various disciplines and interests have been sought to assure that if the introductions are successful, the effects will be beneficial, while if unsuccessful, there will be no negative effect.

REFERENCES – PART IV

1. Andres, L. A. and Bennet, F. D., Biological control of aquatic weeds, *Annu. Rev. Entomol.*, 20, 31, 1975.
2. Bagnall, L. O., Baldwin, J. A., and Hantges, J. E., Processing and storage of water hyacinth silage, *Hyacinth Control J.*, 12, 73, 1974.
3. Bartley, T. R. and Gangstad, E. O., Environmental aspects of aquatic plant control, *ASCE J. Irrig. Drain. Div.*, 100, 231, 1974.
4. Blanchard, J. L., Economic aspects of weed control in the lakes of Winter Park, Florida, *Hyacinth Control J.*, 6, 21, 1967.
5. Brown, A. H., Control of water hyacinth by mechanical means, *Soil Sci. Soc. Fla. Proc.*, 9, 66, 1948.
6. Bryant, C. B., Aquatic weed harvesting – Effects and costs, *Hyacinth Control J.*, 8, 37, 1970.
7. Conway, E. K., Evaluation of *Cercospora rodmani* as a biological control of water hyacinths, *Phytopathology*, 66, 914, 1976.
8. Coulson, J. R., Prognosis for control of water hyacinth by arthropods, *Hyacinth Control J.*, 9, 31, 1971.
9. Cooley, W. W. and Lohnes, R. R., *Multivariate Data Analysis*, John Wiley & Sons, New York, 1970.
10. Couch, R. and Gangstad, E. O., Response of water hyacinth to laser radiation, *Weed Sci.*, 22, 451, 1974.
11. Gangstad, E. O., Aquatic plant control program, *Hyacinth Control J.*, 9, 45, 1971.
12. Gangstad, E. O., Herbicidal control of aquatic plants, *ASCE J. Sanit. Eng. Div.*, 98, 397, 1972.
13. Hitchcock, A. E., Zimmerman, P. W., Kirkpatrick, H., Jr., and Earle, T. T., Growth and reproduction of water hyacinth and alligator weed and their control by means of 2,4-D, *Contrib. Boyce Thompson Inst.*, 16, 91, 1950.
14. Holan, W. J. and Kumse, D. W., The papermaking properties of water hyacinth, *Hyacinth Control J.*, 12, 91, 1974.
15. Hughes, J. S. and Davis, J. T., Variations in toxicity to bluegill sunfish of phenoxy herbicides, *Weeds*, 11, 50, 1963.
16. Inglis, A. and Davis, E. L., The effect of water hardness on the toxicity to fish of several organic and inorganic herbicides, Proceedings, Weed Science Society of America, New Orleans, 1968.
17. Mount, D. I. and Stephen, C. E., A method for establishing acceptable toxicant limits for fish-malathion and the butoxyethanol ester of 2,4-D, *Trans. Am. Fish. Soc.*, 96, 185, 1967.
18. Parra, J. V. and Hortenstine, C. C., Plant nutritional content of some Florida water hyacinths and response by pearl millet to incorporation of water hyacinths in three soil types, *Hyacinth Control J.*, 12, 85, 1974.
19. Perkins, B. D., Potential for water hyacinth management with biological agents, *Proc. Tall Timbers Conf. Ecol. Animal Control Habitat Manage.*, 4, 53, 1973.
20. Rintz, R. E., A zonal leaf spot of water hyacinth caused by *Cephalosporium zonatum*, *Hyacinth Control J.*, 11, 41, 1973.
21. Rodgers, C. and Stalling, D. L., Dynamics of ^{14}C-labeled, 2,4-D butoxyethanol ester in organs of rainbow trout, channel catfish and bluegills, *Weed Sci.*, 20, 101, 1972.
22. Schultz, D. P., Dynamics of a salt of (2,4-dichlorophenoxy)acetic acid in fish, water and hydrosoil, *J. Agric. Food Chem.*, 21, 186, 1973.
23. Schultz, D. P. and Harman, P. D., Residues of 2,4-D in pond waters, mud and fish, *Pestic. Monit. J.*, 8, 173, 1974.
24. Schultz, D. P. and Gangstad, E. O., Dissipation of residues of 2,4-D in water, hydrosoil and fish, *Aquat. Plant Manage. J.*, 14, 43, 1976.
25. Smith, G. E. and Isom, B. G., Investigation of effects of large-scale applications of 2,4-D on aquatic fauna and water quality, *Pestic. Monit. J.*, 1, 16, 1967.
26. Vogel, E. and Oliver, A. D., Jr., Evaluation of *Arzama densa* as an aid in the control of water hyacinth in Louisiana, *J. Econ. Entomol.*, 62, 142, 1969.
27. Vogel, E. and Oliver, A. D., Jr., Life history and some factors affecting the population of *Arzama densa* in Louisiana, *Ann. Entomol. Soc. Am.*, 62, 749, 1969.
28. Walker, C. R., Toxicological effects of herbicides on the fish environment, Proceedings of the Eighth Annual Air and Water Pollution Conference, *Univ. Mo. Bull. Eng. Exp. Stn. Ser.*, 2, 17, 1963.
29. Weldon, L. W. and Blackburn, R. D., Herbicidal treatment effect on carbohydrate levels of alligator weed, *Weed Sci.*, 16, 66, 1968.
30. Whitney, E. W., Montgomery, A. B., Martin, E. C., and Gangstad, E. O., The effects of a 2,4-D application on the biota and water quality in Currituck Sound, North Carolina, *Hyacinth Control J.*, 11, 13, 1973.
31. Wojtalik, T. A., Hall, T. F., and Hill, L. O., Monitoring ecological conditions associated with wide-scale applications of DMA 2,4-D to the aquatic environment, *Pestic. Monit. J.*, 4, 184, 1971.
32. Wunderlich, W. E., The use of machinery in the control of aquatic vegetation, *Hyacinth Control J.*, 6, 22, 1967.
33. Wunderlich, W. E., Aquatic plant control and the dollar, *Hyacinth Control J.*, 7, 28, 1968.
34. Zeiger, C. F., Hyacinth obstructions to navigation, *Hyacinth Control J.*, 1, 16, 1968.
35. Center, T. D. and Balciunas, J., Department of Entomology, University of Florida, unpublished data.
36. Van Dyke, J. M., Mechanical harvesting of water hyacinth in Shell Creek Reservoir, Charlotte County, Florida, in Research Planning Conference on Aquatic Plant Control Project, U.S. Army Corps of Engineers, Waterways Experiment Station, Vicksburg, Miss., 1972, p. E3.
37. Stallings, D. L. and Hutchins, J. N., Fish-Pesticide Research Laboratory, Columbia, Mo., unpublished data.
38. Schutz, D. P., Fish-Pesticide Laboratory, Warm Springs, Ga., unpublished data.

39. **Bock, J. H.,** An ecological study of *Eichhornia crassipes* with special emphasis on reproductive biology, Ph.D. thesis, University of California, Berkeley, 1966.
40. **Silveira-Guido, A.,** Jackson Agricultural School, Montevideo, Uruguay, unpublished data.
41. **Perkins, B. D.,** unpublished data.
42. **Perkins, B. D.,** Biocontrol of water hyacinth, in Aquatic Plant Control Program Tech. Rep. No. 6, Biological Control of Water Hyacinth with Insect Enemies, U.S. Army Corps of Engineers, Waterways Experiment Station, Vicksburg, Miss., 1974, p. E3.
43. **Coulson, J. R.,** Potential environmental effects of the introduction of the Argentine water hyacinth weevil, *Neochetina eichhorniae,* into the United States, in Aquatic Plant Control Tech. Rep. No. 6, Biological Control of Water Hyacinth with Insect Enemies, U.S. Army Corps of Engineers, Waterways Experiment Station, Vicksburg, Miss., 1974, p. F3.

Appendix

GLOSSARY OF ECOLOGICAL TERMS

Abiotic — Nonliving; pertaining to physico-chemical factors only.

Acclimation — Adjustment to environmental change on the part of the individual; the physiological adjustment or increased tolerance shown by an organism to environmental change; nonstandard variant of acclimatization.

Accretion — The process of growth or enlargement by external accumulation.

Acidity — Quality of being acid or sour; having pH less than 7 (see *pH*).

Adaptable — Capable of undergoing inheritable (and/or nonheritable) structural or functional changes.

Adaptation — The result of process of long-term evolutionary adjustment of a population to environmental changes.

Adhesion — The molecular attraction exerted between the surfaces of bodies in contact; steady or firm attachment, usually of an organism to a nonliving surface.

Adsorption — The adherence of substances to the surfaces of bodies with which they are in contact, but not in chemical combination.

Aerobic — Living, active or occurring only in the presence of oxygen or air.

Algae — Any of a group of chiefly marine or freshwater chlorophyll-bearing aquatic plants with no true leaves, stems, or roots. Ranging from microscopic single-cell organisms or colonies to large macroscopic seaweeds, etc.

Alkali — A soluble mineral layer present in quantities detrimental to agriculture in some soils of basic pH in arid regions; a soluble mineral (salt) obtained from the ashes of plants and consisting largely of potassium or sodium carbonate.

Alkalinity — The quality of being alkaline or basic in pH, i.e., greater than pH 7 (the opposite of acid).

Allochthonous — Pertaining to material derived from outside habitat or environment under consideration, e.g., allochthonous detritus of a lake is that derived from surrounding terrestrial environment or from influent streams.

Algal bloom — Rapid and flourishing growth of algae.

Alluvial — Pertaining to *alluvium* (q.v.).

Alluvium — Sediments, usually mineral or inorganic, deposited by running water.

Alpine — That region which is above the montane timberline; characterized by the presence of herbs and grass-like plants and low, slow-growing shrubs.

Ambient — Surrounding on all sides.

Ammonification — Production of ammonia in decomposition of nitrogen-containing compounds such as proteins.

Amphibian — Any of a class of vertebrate (q.v.) animals most of which pass through an aquatic larval stage with gills and then through a terrestrial stage with lungs (e.g., salamanders, frogs and toads.)

Anadromous — Pertaining to fish (such as salmon) which ascend fresh water streams from salt water to spawn.

Anaerobic — Capable of living or active in the absence of air or free oxygen.

Annual plants — A plant which grows from seed and reproduces in one year.

Anoxic — Pertaining to conditions of oxygen deficiency.

Aphotic zone — The lower portion of bodies of water not reached by light.

Aquaculture — Production of food from managed aquatic systems.

Aquatic — Growing, living in, frequenting, or pertaining to marine or fresh water.

Aquatic habitat — A habitat (q.v.) located in water.

Aquifer — A water-bearing stratum of permeable rock, sand, or gravel.

Arctic — Of, or characteristic of, the region around the North Pole to approximately $65°$ N latitude; all regions north of the boreal timber line.

Assimilation — Transformation of absorbed nutrients into body substances.

Association — A definite or characteristic assemblage of plants living together in an area essentially uniform in environmental conditions; any ecological unit of more than one species.

Associes — An association (q.v.) constituting a temporary stage of plant succession; a non-climax community to be replaced by another in the process of succession (q.v.).

Autotrophic — The nutrition of those plants that are able to construct organic matter from inorganic.

Autotrophs – Organisms capable of autotrophic (q.v.) growth. See also producers.

Bacteria – Any of a class of free-living, parasitic, or pathogenic microscopic organisms having single-celled or noncellular bodies; devoid of conventional nuclei; often living in colonies; colloquially known as microbes.

Baseline profiles – Used for a complete survey of the environmental conditions and organisms existing in a region prior to unnatural disturbances.

Beach – Depositional area at the shore of an ocean or lake covered by mud, sand, gravel, or larger rock fragments and extending into the water for some distance.

Benthic – Pertaining to the bottom of lakes or oceans, or to organisms which live on the bottom of water bodies.

Benthos – Those organisms which live on the bottom of a body of water.

Bioassay – Determination of the physiological effect of a substance (such as a drug) by comparing its effects on a test organ or living organism with that of some standard substance; in contrast to chemical assay or analysis.

Biocoenosis – An ecological unit comprising both the plant and animal populations of a habitat; a biological or biotic community.

Biocide – Any chemical or agent that kills organisms.

Biodegradable – Can be broken down to simple inorganic substances by the action of decay organisms (bacteria or fungi).

Biological diversity – The number of kinds of organisms per unit area or volume; the richness of species in a given area.

Biochemical oxygen demand – The amount of oxygen required to decompose (oxidize) a given amount of organic compounds to simple, stable substances.

Biological concentration – The active concentration of a substance (molecule or compound) by an organism as a result of normal activities, e.g., absorption or ingestion.

Biological magnification – The step by step concentration of substances in successive levels of food chains (q.v.); commonly reported only for harmful substances.

Biological processes – Processes characteristic of, or resulting from, the activities of living organisms.

Biological productivity – See *productivity*.

Biological stability – See *stability*

Biogeochemical cycling – The movement of chemical elements from the physical environment to organisms in an ecosystem and back to the environment.

Biomass – The total weight of matter incorporated into (living and dead) organisms.

Biome – Any of the major terrestrial ecosystems of the world such as tundra, deciduous forest, desert, taiga, etc.

Biosphere – That portion of the solid and liquid earth and its atmosphere where living organisms can be and are sustained; "organic nature" in general.

Biota – All of the named or namable organisms of an area; fauna and flora (= biota) of a region.

Biotic – Of life.

Biotic succession – See *succession.*

Biotic potential – The inherent ability of members of a population to grow in numbers within a given time and under stated environmental conditions.

Biotic community – See *community;* biotic implies plants and animals.

Biotic pyramid – The set of all food chains (q.v.) or hierarchic arrangements of organisms as eaters and eaten in a prescribed area when tabulated by numbers or by biomasses; usually takes a pyramidal form.

Biotope – A segment, usually a small segment, of a habitat (q.v.).

Biotype – A genetically homogeneous population composed only of closely similar individuals; a genotypic race or group of organisms.

Buffer – An intermediate region or ecotone (q.v.) between two systems whose presence reduces the effects of one system on the other; a chemical substance which tends to maintain constant pH.

Bivalve – A gastropod having a shell composed of two valves, e.g., a clam.

Bloom – To flower; of algae, to appear o occur suddenly or in large quantity or degree; se *algal bloom.*

Bog – A quagmire or wet, spongy ground often a filled-in lake; composed primarily of dea plant tissues (peat), principally mosses.

Brackish water – Water, salty between th concentrations of fresh water and sea wate usually 5 to 10 parts per thousand.

Calorie – (Abbreviated cal.) The heat require (at 1 atm) to raise the temperature of 1 g of wat $1°C$ (specifically from 4 to $5°C$. A kilocalor (abbreviated kcal) in nutrition, is 1000 calories.

Canopy — The leafy cover of vegetation, e.g., the uppermost leafy layer in forests.

Carnivore — An organism that eats living animals.

Carcinogen — A substance or agent producing or inciting cancer.

Carrying capacity — The maximum population size of a given species in an area beyond which no significant increase can occur without damage occurring to the area.

Catadramous — Living in fresh water and going to salt water to spawn.

Catharobes — The organisms of "pure" water, poor in organic matter.

Chemical oxygen demand — See *Biochemical Oxygen Demand.*

Chemosynthesis — The process by which some organisms (bacteria) obtain their energy for CO_2 assimilation by the chemical oxidation of simple inorganic compounds (in contrast to photosynthesis).

Chlorophyll — The green, photosynthetic pigments of plants.

Climate — The average conditions of the weather over a number of years; macroclimate is the climate representative of relatively large area; microclimate is the climate of a small area, particularly that of the living space of a certain species, group, or community.

Climax — The final, stable community in an ecological succession (q.v.) which is able to reproduce itself indefinitely under existing conditions.

Climax community — See *climax.*

Cline — A continuous series of differences (structural or functional) exhibited by a group of related organisms, usually along a line of geographic or environmental gradient.

Closed system — An organized assemblage of system objects, in which there is not exchange of material with objects outside of the system.

Clone — The vegetatively produced, genetically identical, offspring of a single individual.

Codominant — Any of equal dominant forms; one of several species which dominate a community, no one to the exclusion of the others.

Coliform — Structurally and functionally resembling certain bacteria of the vertebrate intestine called *Bacillus* or *Escherichia coli.*

Coliform Levels — Numbers referring to the density of coliform bacteria in water bodies.

Colloid — A dispersion of particles larger than small molecules and which do not pass through semipermeable membranes and do not settle out.

Colonizing — Pertaining to organisms which occupy areas previously barren, or at least areas presently unoccupied by that species.

Community — All of the plants and animals in an area or volume; a complex association usually containing both animals and plants.

Community metabolism — The combined metabolism (metabolic activity) of all organisms in a given area or community.

Community respiration — The combined respiration of all organisms in a community.

Compensation level — Depth in a body of water at which the light available is just sufficient to allow enough photosynthesis to balance respiration over an appreciable time.

Competition — Interaction of organisms which utilize common resources in short supply.

Conjugation — The fusion of two similar (i.e., not obviously differentiated) gametes; usually in contrast to fertilization (q.v.).

Conservation — Supervision, management, and maintenance of natural resources.

Consocial — (rare) Pertaining to plant species found together in a given community.

Consumer — An organism that consumes another.

Consumer (primary) — An organism which consumes green plants.

Consumer (secondary) — An organism which consumes a primary consumer (q.v.).

Copepods — A large subclass of usually minute (0.1 to 5 mm.), mostly pelagic or free-swimming fresh- or saltwater crustaceans (q.v.).

Cover (ground cover) — Vegetation used to reduce wind and water erosion of bare soil.

Crustacean — A large class of (arthropodan) animals, usually aquatic, bearing a horny shell, such as lobsters, barnacles, shrimps, water fleas, etc.

Cryptogam — A plant which does not produce flowers or true seeds: ferns, mosses, liverworts, and algae.

Cybernetics — The study and design of feedback control systems.

Cyst — A pouch or sac without an opening, such as a resting spore in certain algae, or bacteria or a parasite walled-off within the body of the host.

Deciduous — Falling off or actively shed at maturity or at certain seasons.

Decomposers — Those organisms, usually bacteria (q.v.) or fungi, which participate in the breakdown of large molecules associated with organisms. Hence, those organisms which recycle dead organisms.

Delta — The alluvial deposit at the mouth of a river.

Demersal — Applied to eggs which are heavy and sink to the bottom of a stream or other body of water.

Denitrification — Chemical conversion of nitrates to molecular (gaseous) nitrogen (N_2) or to nitrous oxide or to ammonia by bacteria or by lightning.

Desiccation — Drying out; a method by which organisms and their disseminules (q.v.) survive unfavorable periods.

Detritus — A nondissolved product of disintegration or wearing away. Pertains to organic or inorganic matter.

Devonian — Pertaining to that period of geologic time (or the rocks of that period) marked especially by the major evolution of aquatic vertebrates (q.v.); the "age of fish."

Diatom — Any of a class of minute, planktonic or attached unicellular or colonial algae with cases of silica (opal).

Disseminule — General term for seeds, spores, resting eggs, pelagic eggs, or larvae, etc.

Dissolved oxygen — An amount of gaseous oxygen dissolved in volume of water.

Disclimax — A climax (q.v.) which is the consequence of repeated or continuous disturbance by man, domesticated animals, or natural events.

Distribution — The geographic range of a species.

Diurnal — Pertaining to phenomena of daily occurrence; of that portion of the day in which light occurs.

Dominance — The degree of influence (usually inferred from the amount of area covered) that a species exerts over a community.

Dominant — An organism that controls the habitat at any stage of development; in practice the organism that is most conspicuous and covers the most area.

Dune — A hill of drifting sand usually formed on existing or former shores or coasts, but often carried far inland by prevailing winds.

Dynamic Equilibrium — A state of relative balance between forces or processes having opposite effects.

Dystrophic — a type of lake in which the water usually has an acid reaction and brown peaty color; lack of nutrients; often associated with bogs (q.v.).

Ebullition — The act, process, or state of boiling or bubbling up.

Ecocline — Gradual changes in the morphological or physiological features in organisms along an environmental gradient.

Ecology — The study of the interrelationships of organisms with and within their environment.

Ecological amplitude — Pertains to the breadth of a species' tolerance (q.v.) to an environmental factor.

Ecological balance — Range of response normally expressed by an unperturbed ecosystem.

Ecological conscience — As defined by Aldo Leopold, a philosophical and political concern for conservation of all natural resources, but especially for scenic and other nonobvious values of undisturbed ecosystems.

Ecological Dominance — Pertains to a species' control, competitiveness, and alteration of conditions for the remainder of the community (q.v. species dominant).

Ecological equivalent — Analogous species in similar environmental contexts; that is, distantly related species displaying closely similar adaptive mechanisms, like loons and cormorants, flying squirrels and flying phalangers, etc.

Ecological indicator — Use of certain species tolerances (q.v.) to reflect or infer more general environmental characteristics; see *indicator*.

Ecological niche — The functions of the organism in its ecological setting; see *niche*.

Ecological resilience — A system's ability to return to a prior state following environmental perturbation (stress).

Ecological succession — See *succession*.

Ecological system — See *ecosystem*.

Ecosystem — A community and its (living and nonliving) environment considered collectively; the fundamental unit in ecology. May be quite small, as the ecosystem of one-celled plants, in drop of water, or indefinitely large, as in the grassland ecosystem.

Ecosystem analysis — Examination of structure, function, and control mechanisms present and operating in an ecosystem.

Ecosystem dynamics — Characteristic and measurable processes within an ecosystem such (1) succession (q.v.), (2) energy flow and nutrient cycling (q.v.), or (3) community metabolism.

Ecosystem function – Energy flow and material production cycling within an ecosystem.

Ecosystem integrity – Implications of ecosystem properties as a whole, especially of resilience (q.v.) or its lack.

Ecosystem structure – The who, what, and where of an ecosystem; its functionally important and weighable components, mostly organisms; the pattern of organism's interrelations and spatial arrangements.

Ecosphere – Envelope of the earth's surface where biological and ecological activities occur. See *biosphere*.

Ecotone – A transition zone between two recognized communities (q.v.).

Ecotype – Race or subdivision of species adapted to local habitat and climate. These genetic groups are broader than a biotype (q.v.) and narrower than a species. Ecocline (q.v.).

Edaphic – Pertaining to the influence of the soil, especially on the plants growing upon it.

Edaphic climax – Self-perpetuating community where soil is limiting further succession at a stage believed to be short of climatic potential.

Edge effect – Phenomena such as changed diversity and/or density of organisms that occur in the vicinity of community boundaries (ecotones, q.v.).

Efficiency (ecological) – Defined exchange of energy and/or nutrients between trophic (q.v.) levels; usually the ratio between production (q.v.) of one level and that of a lower level in the same food chain (q.v.).

Embryo – An early stage of development of animals or plants. Usually experienced by an egg after fertilization and before "hatching."

Emergent – Aquatic plants, usually rooted, which during part of their life cycle have portions above water.

Endangered species – Species that are in danger of becoming extinct.

Endemic – Indigenous or native in a restricted locality; confined naturally to a certain limited area or region.

Endotherms – "Warm-blooded" animals. Animals which have the facility to regulate their body temperatures over a wide range of external temperatures (see *homeotherms*).

Energy (ecological) – Most commonly, that portion of the visible solar radiation (light) captured by plants and ultimately used for food by the animals in an ecosystem.

Energy budget – A quantitative account sheet of inputs, transformations, and outputs of energy in an ecosystem. May apply to the long-wave radiation (heat) of an organism or a lake, or to the food taken in and subsequently reduced to heat by an individual or a population.

Energy cycling – Although this term is sometimes used to imply that the ecological energy in an ecosystem is reused, the term is incorrect. Use instead energy flow.

Energy flow – The one-way passage of energy (largely chemical) through a system, entering via photosynthesis, being exchanged through feeding interactions, and at each stage, being reduced to heat.

Energy subsidy – The man-induced addition of energy designed to increase the production by ecological energy, for example, the use of fossil-fuel energy in tractors to increase the amount of solar energy available from agricultural crops.

Energy system (in ecosystem) – The ecosystem carries out a number of functions including energy flow and the cycling of numerous elements and materials. This energy flow, including the energetic equivalent of the materials, is the energy system of the ecosystem.

Energy transfer process – Any process which transfers energy from one component in an ecosystem to another. Photosynthesis, feeding, and bacterial breakdown are examples.

Enteric – Pertaining to the gut or digestive tract.

Entire – (morphology) Individual or linear in outline, as an entire (non-toothed) leaf margin.

Entropy – The state of thermal disorganization of a system. In a system, entropy is proportional to the nonusable heat produced.

Environment – The sum total or the resultant of all the external conditions which act upon an organism.

Environmental amenities – Attractive or esthetically pleasing environments or portions of environments.

Environmental criteria – Standards of physical, chemical, and biological (but sometimes including social, aesthetic, etc.) components that define a given quality of an environment.

Environmental effect – Resultant of natural or man-made perturbations of the physical, chemical, or biological components making up the environment.

Environmental inventory – A listing of the

components making up an environment, or a listing of types of environments.

Environmental monitoring — The systematic (simultaneous or sequential) measuring of various components constituting the environment.

Environmental quality — Human (individual or social) considerations of desirable ecological situations.

Environmental resistance — The restrictions imposed upon the numerical increase of a species by the physical and biological factors of the environment.

Environmental science — All sciences contributing to understanding of the total environment.

Environmental setting — Environmental context.

Environmental stress — Perturbations likely to cause observable changes in ecosystems; usually departures from normal or optimum. See *stress*.

Environs — The neighborhood; surroundings.

Enzyme — An organic catalyst of protein nature.

Epibenthos — Life forms attached to and growing upon rather than within the bottoms of standing and flowing waters.

Epilimnion — The turbulent superficial layer of a lake between the surface and a horizontal plane marked by the maximum gradient of temperature and density change. Above the hypolimnion (q.v.).

Epiphytes — Plants which grow on other plants but which are not parasitic.

Epizootic — Any organism causing disease in many animals of one kind at the same time; an animal epidemic (epidemic meaning "upon the people").

Equilibrium — In environmental science, a steady state in a dynamic system, with outflow balancing inflow, about which the system ordinarily fluctuates to some small degree. (Often applied to an animal population at zero growth, to the steady interaction of two species [predator-prey], to the energy flow through an ecosystem, and to the nutrient cycling pattern of an ecosystem.)

Erosion — The removal of soil or rock by wearing away of land surface.

Estuarine — Of the mouth region of a river that is affected by tides.

Ethnological — Of that branch of anthropology that deals with extant races and cultures ("peoples"), rather than with language or with extinct cultures.

Ethology — Study of behavior of organisms usually or preferably in their natural environment.

Euphotic — Of the upper layers of water in which sufficient light penetrates to permit growth of green plants.

Euryhaline — Pertaining to species peculiar to or living in brackish waters of marine or non-marine origin, and which are resistant to great changes in salinity.

Eurythermal — Pertaining to organisms having the ability of living through a wide range of temperature conditions.

Euryptopic — Adaptation of species to widely varied conditions (places).

Eutrophic — (literally, "well-fed") Of lakes characterized by the paucity or absence of oxygen in the bottom waters, as a consequence of high primary production and high nutrient content.

Exotherms — Organisms like fish, reptiles, and insects which cannot regulate their own body temperature independent of the temperature of their surroundings.

Exotic — Of any nonnative or rare species; usually introduced.

Excretion — Elimination of waste material from the body of an organism.

Faciation — (rare) A portion of an association (q.v.) in which one or more of the dominants have dropped out and been replaced by other forms, the general community aspect remaining unchanged.

Fauna — The animals of a given region taken collectively; as in the taxonomic sense, the species, or kinds, of animals in a region.

Fecundity — Capacity to produce offspring; in insect ecology, the number of eggs per female that hatch or become larvae.

Feedback — Principle of information returning to sender or to input channel, thus affecting output.

Fertilization — Sexual union at the cellular level; fusion of the nuclei of a male and a female gamete.

Fetch — The expanse of open water which can be affected by the wind.

Fidelity — The degree to which species are confined to certain communities.

Fish kill — Pertaining to sudden death of fish population.

Fishery — Pertaining to fish populations as the basis of an industry, recreational or commercial.

Flood plain — That portion of a river valley which is covered in periods of high (flood) water ordinarily populated by organisms not greatly harmed by short immersions.

Flora — Plants; organisms of the plant kingdom; specifically, the plants growing in a geographic area, as the flora of Illinois.

Flora (micro) — Usually bacteria or fungi.

Fluctuation — Change in level.

Fluvial — Applied to plants growing on streams.

Food chain — Animals linked together by food and all dependent, in the long run, on plants.

Food web — See *food chain*. Implies many cross connections rather than straight line connections.

Forage — To search for, pursue, capture, and ingest food.

Forage fish — Fish eaten by other fish.

Forb — An herbaceous plant, not a grass.

Forest — An association (q.v.) dominated by trees; usually defined as woody plants over 10 m tall.

Forest cover type — All trees and other woody plants (underbrush) covering the ground in a forest. Includes trees, herbs and shrubs, litter, and the rich humus of partly decayed vegetable matter at the surface of the soil. See *forest type*.

Fragile — Easily broken or disrupted; usually refers to communities or ecosystem particuarly susceptible to disruption by man.

Freshet — An overflowing of a stream swollen by heavy rain or melted snow, usually occurring in the spring.

Genetics — The study of heredity or inherited features in individuals or populations.

Genus — A unit of biological classification (taxonomy, q.v.) including one or several species sharing certain fundamental characteristics, supposedly by common descent.

Glade — An open space in a woods or forest.

Greenbelt — A plot of vegetated land separating or surrounding areas of intensive residential or industrial use and devoted to recreation or park uses.

Gradient — A more or less continuous change of some property in space. Gradients of environmental properties are ordinarily reflected in gradients of biota.

Greenhouse effect — Warming of the earth's surface resulting from the capacity of the atmosphere to transmit short-wave energy (visible and ultra violet light) to the earth's surface, and to absorb and retain heat radiating from the surface; carbon dioxide and water vapor in the atmosphere both contribute to the effect.

Ground water — Water found underground in porous rock or soil strata.

Grassland — An area in which grasses are the major plants; trees and shrubs are largely absent.

Habitat — The environment, usually the natural environment in which a population of plants or animals occurs.

Habitat structure — The physical structure of a habitat; e.g., the layering of vegetation in a forest, or the grain of a coral reef.

Halophyte — A plant which grows in salty soil or water.

Halophytic — See halophyte.

Heat budget — The quantitative listing of all heat inputs, transformations, and outputs of an ecosystem or organism.

Herbaceous — Pertaining to any plant lacking woody tissue in which the leaves and stem fall to ground level during freezing or drying weather.

Herbicide — A chemical substance used to kill plants or inhibit plant growth.

Herbivore — An organism which eats living plants or plant parts (e.g., seeds).

Herpetofauna — The amphibian and reptile species characteristic of an area.

Heterotrophs — Organisms which must obtain their food from living or dead organic matter.

Higher plants — "Flowering" plants which reproduce by seeds; phanerogams, not cryptogams (q.v.).

Holocoenotic — Pertaining to a system which is organized so that the total system has properties not present in its individual parts; as this is true of all systems, the term is superfluous but often used for emphasis. See *ecosystem*.

Holomictic — A lake in which surface and bottom waters are completely mixed by vertical circulation, occasionally at least.

Home range — The area or space of normal activity of an individual animal; sometimes, but not necessarily defended against intrusion by other individuals. See *territory*.

Homeostasis — The inherent stability or self-regulation of a biological system; the ability of such a system to resist external changes.

Homeotherms — Animals (mammals and birds) which maintain a more or less constant body temperature despite variations in external temperature; warm-blooded animals or endotherms (q.v.).

Humic — Pertaining to the soil- or water-borne substance resulting from the partial decay of leaves and other plant material.

Hydrarch — Pertaining to successions which originate in aquatic habitats such as lakes and ponds and progress toward more terrestial conditions, as in bogs and swamps.

Hydric — Characterized by or pertaining to conditions of abundant moisture supply.

Hydrodynamics — The branch of physics which studies the movements of water and other liquids.

Hydrophyte — A plant which grows in water or very wet earth.

Hypolimnion — In certain lakes, the portion (below the zone of warmer water) which receives no heat directly from sunlight and no aeration by vertical circulation.

Indicator — An organism or collection of organisms having relatively narrow requirements and thus indicating the presence of certain environmental conditions.

Indicator organism — See *indicator*

Indigenous — Pertaining to native species; not introduced.

Influent — Anything flowing into something, as a stream to a lake; (rare or obsolete) an organism such as a parasite which has important but nonobvious relations in the biotic balance and interaction.

Inhibitor — A substance which slows or prevents growth or reproduction of an organism.

Inorganic — Pertaining to matter that is neither living nor immediately derived from living matter; typically does not contain carbon, but carbon dioxide is inorganic.

— **Insecticide** — Any substance used to kill insects.

Interaction — A relationship in which each component influences the other.

Interface — A surface that lies between two areas or volumes and forms their common boundary.

Intertidal — Pertaining to the region of marine shoreline between high-tide mark and low-tide mark; where neap, spring, and storm tides are important; usage is flexible.

Inversion — In meteorology, a condition in which cooler surface air is trapped under an upper layer of warmer air, preventing vertical circulation.

Invertebrate — Any animal lacking a backbone (i.e., insects, spiders, crustaceans, segmented worms, round and flatworms, molluscs, etc.)

Irrigation — To supply agricultural land with water by artificial means.

Isohaline — In an organism, the property of being in balance with the salt concentration of its surroundings.

Isotherm — A line on a map connecting points having the same temperature at the same time.

Keystone species — A species whose removal causes marked changes to the community.

Lacustrine — Originating in, or inhabiting, a lake.

Lagoon — A shallow area of water generally separated from a larger body of water by a partial barrier.

Lag time — The delay between some event and its effect.

Lake — A large body of water contained in a depression of the earth's surface and supplied from drainage of a larger area. Locally may be called a pond.

Lake turnover — The complete top-to-bottom circulation of water in a lake which occurs when the density of the surface water is the same or slightly greater than that at the lake bottom; most temperate zone lakes circulate in spring and again in fall.

Larva — An early developmental stage of an animal which changes structurally when it becomes an adult (e.g., a tadpole or caterpillar).

Leachate — The soluble material which is washed or dissolved during leaching.

Leaching — The removal of various soluble materials from surface soil layers by the passage of water through (around) the layers.

Lentic — Pertaining to still or slowly flowing water situations (e.g., lakes, ponds, swamps).

Life cycle or life history — The series of changes or stages undergone by an organism from fertilization, birth, or hatching to reproduction of the next generation.

Life system — (rare) A population and the environment which influences it; see ecosystem.

Life zone — (rare) An altitudinal or latitudinal zone defined by climatic characteristics and having certain plants and animals (especially birds and mammals).

Limiting factor — An environmental factor (or factors) which limits the distribution and/or abundance of an organism or its population, i.e., the factor which is closest to the physiological limits of tolerance of that organism.

Limnetic zone — The open water zone of a lake or pond from the surface to the depth of effective light penetration; offshore, from areas shallow enough to support rooted aquatic plants.

Limnology — The study of the biological, chemical, and physical features of inland waters.

Littoral — Pertaining to the shoreward region o

a body of water in which light penetrates to the bottom; in lakes or ponds, from shoreline to the lakeward limit of rooted aquatic plants; in oceans, from shoreline to a depth of 200 m.

Lotic – Pertaining to rapid water situations; living in waves or currents.

Macrofauna – The large (i.e., visible to the naked eye) animals of an area.

Macrophytes – Large-bodied aquatic plants; nonmicroscopic.

Marine – Pertaining to the sea or ocean.

Marsh – A tract of low-lying, soft, wet land, commonly covered (sometimes seasonally) entirely or partially with water; a swamp dominated by grasses or grass-like vegetation.

Meristic variation – Variation among segments in a segmented animal.

Meromictic – Lakes which undergo only a partial circulation.

Mesic – Characterized or pertaining to conditions of medium moisture supply.

Metalimnion – The zone over which temperature drops relatively rapidly with depth. See thermocline.

Microbiota – The microscopic organisms present in an area or volume.

Microclimate – Conditions of moisture, temperature, etc., as influenced by the topography, vegetation, and the like. See *climate*.

Microenvironment (habitat) – A small or restricted set of distinctive environmental conditions, such as a dead animal or a fallen log.

Microflora – The microscopic plants present in an area or volume.

Microphyte – The smaller algae, e.g., diatoms.

Migration – A regular movement from one region to another.

Minimal area – The smallest area upon which a community reaches its mature or developed stage, including all of its characteristic components.

Mixotrophic – Fed by several alternative modes of nutrition; usual for some one-celled animals and plants.

Mollusc – Any of a phylum of invertebrate animals including oysters, clams, mussels, snails, slugs, squids, octopi, whelks, and other shellfish.

Monoculture – Cultivation of land in a single crop.

Monomictic – A polar or tropical lake in which the water never exceeds or falls below (respectively) 4°C, and thus has only one period of turnover or circulation per year.

Monospecific – Pertaining to a single species of organism.

Morbidity – In medical ecology, the incidence (measured frequency) of disease in a population; the illness rate.

Morphology – The study of the form and structure (but not the functions) of an organism.

Mortality – Death in a population; the death rate.

Mosaic – A patchwork pattern of distribution of habitats or communities.

Muskeg – Moss-covered countryside, or continuous boggy ground; e.g., moss bogs of the Canadian forest.

Mutualism – A form of interrelationship between two organisms in which both involved organisms benefit. See *symbiosis.*

Nanoplankton – Extremely small, free-floating aquatic organisms. See *plankton.*

Natality – An expression of the birthrate of a population.

Natural area – An area in which natural processes predominate, fluctuations in numbers of organisms are allowed free play, and human intervention is minimal.

Natural environment – The complex of atmospheric, geological, and biological characteristics found in an area in the absence of artifacts or influences of a well-developed technological human culture; an environment in which human impact is not controlling, or significantly greater than that of other animals. See *natural area.*

Natural pollution – The production and emission by geological or nonhuman biological processes of substances commonly associated with human activities (e.g., natural oil seeps, hydrocarbons or toxins released by plants or animals).

Natural selection – A biological process resulting in differential survival of different gene combinations selected in a particular environment.

Natural setting – The complex of atmospheric, geological, and biological characteristics of an area as they determine its appearance. (See *natural environment.*)

Nekton – Free-swimming organisms of open water, large and strong enough to be independent of turbulent water movement (fish).

Net production – See *productivity, net primary.*

Nertic zone – The zone of shallow water adjoining a coast line.

Neuston – Microorganisms in contact with or in the surface film of water.

Niche — The range of sets of environmental conditions which an organism's behavioral morphological and physiological adaptations enable it to occupy; the role an organism plays in the functioning of a natural system, in contrast to habitat.

Nitrogen fixation — A step in the nitrogen cycle involving hydrogenation (reduction) of molecular nitrogen (N_2) to amino or ammonia nitrogen (NH_2 or NH_3) performed by certain nitrogen-fixing (soil) bacteria and blue-green algae.

Nitrogen gas supersaturation — An excess of dissolved nitrogen which may be toxic to animals, and which causes the "bends" in divers.

Nitrification — A step in the nitrogen cycle technically involving oxidation of nitrogen, e.g., NH_3 from ammonia to nitrates (NO_3). Soil-dwelling (chemosynthetic) bacteria nitrify ammonia in two steps, to nitrite (NO_2) and to nitrate (NO_3) in which form it is most available to plants. Chemical reduction of nitrogen, as to N_2, is denitrification.

Nocturnal — Occurring or active during the period between sunset and sunrise.

Nonrenewable resource — A natural, normally nonliving, resource such as a mineral which is present in finite supply and is not reviewed by natural system.

Nonvascular plants — Plants without specialized conductive tissues (xylem or phloem) e.g., algae, mosses, liverworts.

Nursery area — An area where animals congregate for giving birth or where the early life history stages develop, e.g., estuaries for shrimp, Scammon's lagoon; Baja California, for gray whale.

Nutrients — Chemical elements essential to life. Macronutrients are those of major importance required in relatively large quantities (C, H, O, N, S, and P); micronutrients are also important but required in smaller quantities (Fe, Mo).

Nutrient cycling — The movement of nutrients from the nonliving (abiotic) through the living (biotic) parts of the environment and back to the abiotic parts.

Nymphs — Immature stage of Arthopods (primarily insects) with incomplete metamorphosis; not larva because not sufficiently different from adults.

Oligoaerobe — Organism which thrives at low oxygen concentrations.

Oligoaerobic — Conditions of low oxygen "tension" (pressure or concentration).

Oligotrophic — (lit. "poorly fed") Of lakes characterized by abundant oxygen in deep water as a consequence of small nutrient supply and low productivity of organic material; see *eutrophic*.

Omnivorous — Eating a wide variety of food, both plant and animal.

Ontogeny — Development of an individual organism.

Organic — Compounds containing carbon [and hydrogen]; living or derived from living matter.

Organic detritus — Particles or fragments of a larger living or recently dead body produced by its disintegration. In aquatic systems, finely divided, settleable particles whose continued destruction consumes oxygen.

Organism — Any living or recently dead thing.

Overturn — The complete circulation or mixing of the upper and lower waters of a lake when the temperatures (and densities) are similar.

Oxygen depletion — Removal or exhaustion of oxygen by chemical or biological use.

Oxygen sag — A drop in oxygen concentration, usually at night, due to respiration.

Parameter — A measurable, variable quantity as distinct from a statistic or estimate.

Parasite — An organism living on or in a living organism, without killing it immediately, and deriving its nutrition from it with a detrimental effect on the host.

Passerine — Perching birds (e.g., sparrow) including all songbirds.

Pelagic zone — Free open water of the ocean or lake with no association with the bottom.

Periodicity — A regular cyclic behavior of an organ, cell, or organism in time (see *photoperiodism*).

Periphyton — Community of organisms, usually small but densely set, closely attached to stems and leaves of rooted aquatic plants or other surfaces projecting above the bottom.

Permafrost — Permanently frozen subsoil, thawing at the surface in summer, characteristic of Arctic tundra.

Pesticide — Toxic chemical used for killing organisms. Usually widely toxic to living things (see *herbicide, insecticide*).

pH — ("power hydrogen") Negative logarithm of hydrogen-ion concentration, a numerical expression of acidity (see *acid, alkaline*).

Oxidation — A reaction between molecule involving transfer of an electron-from a reduced to oxidized molecule (see *reduction*); ordinarily in volves gain of oxygen and/or loss of hydrogen, i.e., dehydrogenation.

Phenology — The study of the periodic phenomena of nature, especially animal and plant life in their relations to weather and climate (e.g., bird migration, flowering, bud opening, freezing and thawing).

Photoperiodism — Response of plants and animals to relative duration of light and darkness.

Phototropism — A growth curvature of a plant in response to a unilateral light source; (obsolete) behavioral response of an animal or microbe to light stimulus.

Poikilothermic — A "cold-blooded" organism whose body temperature varies approximately with the environment. Generally other than birds and mammals.

Pollutant — A residue (usually of human activity) which has an undesirable effect upon the environment (particularly of concern when in excess of the natural capacity of the environment to render it innocuous).

Pollution — An undesirable change in atmospheric, land, or water conditions harmfully affecting the material or aesthetic attributes of the environment.

Polyclimax — Two or more simultaneously existing, stable, self-maintaining communitites controlled by local environmental conditions in a larger area (see *climax*).

Pond — A small lake.

Population — A group of organisms of the same species.

Population density — The number of individuals of a population per unit area, or volume.

Population index — An estimate of size or other characteristic of a population, obtained by indirect means (e.g., by songs, droppings).

Population irruption — A sudden, large increase in population density, resulting in emigration or immigration.

Population pressure — A metaphor implying the magnitude of demand of a population on space or other resources.

Potamology — The study of streams, especially large rivers.

Polythermal — Confined to high temperatures (rare); in contrast to oligothermal (g.v.).

Predator — An organism, usually an animal, which kills and consumes another organism in whole or part.

Predator chain — See *food chain;* also *trophic* and *biotic pyramid.*

Prey — An organism killed and at least partially consumed by a predator.

Predominant (= dominant) — An organism of outstanding abundance or obvious importance in a community (q.v.).

Pristine state — A state of nature without human effect or with negligible human effect.

Producer (= producer organism) — An organism which can synthesize organic material using inorganic materials and an external energy source (light or chemical). See *autotroph;* also *biotic pyramid.*

Production — The amount of organic material produced by biological activity in an area or volume.

Productivity — The rate of production of organic matter produced by biological activity in an area or volume (e.g., grams per square meter per day, or other units of weight or energy per area or volume and time).

Productivity, gross primary — The rate of synthesis of organic material produced by photosynthesis (or chemosythesis), including that which is used up in respiration by the producer organism.

Productivity, net primary — The rate of accumulation of organic material in plant tissues. Gross primary productivity less respiratory utilization by the producer organism.

Productivity, secondary — The rate of production of organic materials by consumer organisms (animals) which eat plants (which are the primary producers). See *heterotrophs.*

Profundal zone — The bottom of a body of water below the metalimnion (q.v.), or below the limit of macrophytic vegetation (e.g., rooted plants or seaweeds, large algae such as Chara, or mosses).

Prolific — Producing numerous young or fruit; marked by abundant productivity.

Provenance — The geographical source or place of origin of something, e.g., a genetic stock or a lot of seed.

Range — The geographic area of occurrence of a species; the region over which a given form occurs, naturally or after introduction.

Raptors — Any of several birds of prey (hawks, falcons, eagles, owls).

Recent — Informal (geological), usually referring to the period of time from the last glaciation to the present. The United States Geological Survey uses it formally, but has not defined it; most geologists prefer "Holocene."

Recycling — The repeated use of a finite body of resources such as minerals.

Red tide — A reddish color of near-shore marine waters due to the presence of extremely large numbers of red-pigmented microorganisms, which liberate toxins lethal to fish.

Reducers — See *reducer organisms*.

Reducer organisms — Those organisms which have the capability of promoting chemical reductions (see below), as green plants reduce CO_2 and sulfate-reducing bacteria reduce sulfate ($SO_4^=$).

Reduction — A reaction between molecules in which the transfer of an electron is involved (the reduced molecule acquires an electron). Ordinarily involves loss of oxygen and/or addition of hydrogen.

Relict — A species properly belonging to an earlier community type than that in which it is found. A community (or fragment of one) that has survived some important change, and now seems to be or is "left behind."

Relief — Variations in elevation of the earth's surface.

Remote sensing — A method for determining the characteristics of an object, organism, or community from afar.

Reptiles — One of the major groups of vertebrate animals, including crocodilians, turtles, lizards, and snakes, having scales or horny plates, true lungs, and a three- or four-chambered heart.

Reservoir — An artificially impounded body of water; also, the supply of any commodity, as a reservoir of infection, etc.

Resident — Normally to be found when looked for; of birds, nonmigratory.

Resilience — The ability of any system, e.g., an ecosystem, to resist or to recover from stress.

Resistant — Said of organisms not overly susceptible to environmental stresses; most pests are pestiferous because resistant.

Respiration — (Commonly) breathing; (in biology) the oxidative breakdown of food molecules by cells with the release of energy.

Retrogressing — Changing in a reverse order to a simpler or earlier state.

Reverse osmosis — The movement of water through a semi-permeable membrane in the direction of a concentration gradient; with suitable membranes and energy supplies, the process can be used to purify contaminated water.

Rheotropism — The behavioral response of an organism, cell, or organ to a current of water.

Riverbank overstory — Those plants growing along streams whose canopies occupy the greatest heights.

Riverine — Pertaining to rivers.

Rookery — The breeding or nesting place of colonies of birds, seals, etc.

Rough fish — A non-sport fish, usually omnivorous in food habits, but not prized owing to poor flavor, excessively bony flesh, or inadequate cooperation with anglers.

Ruderal — A weed; an introduced plant species growing under disturbed conditions, in waste places or among rubbish.

Salinity — The concentration of any salt; concentration of sodium chloride is, technically, halinity or sodium chlorinity.

Salinity wedge — The movement of subsurface saline water into an aquifer, or in an estuary. Of a body of saline (sea) water under the fresh water.

Salmonid — Pertaining to salmon, trout, char, and allied freshwater and anadromous fishes.

Salt marsh — Similar to a fresh (grass-dominated) marsh, but adjacent to marine areas covered periodically (tidally or seasonally) with saline water.

Sanctuary — An area usually set aside by legislation or deed restrictions for the preservation and protection of organisms.

Sapropel — Ooze; slimy black or brown sediment of marine, estuarine, or (rarely) lacustrine deposition consisting largely of organic debris. Finely divided, rich in iron and sulfide; and chemically strongly reducing.

Saprobic — Pertaining to forms of life living in foul, badly polluted, or septic waters.

Saprophyte — A plant deriving all its nourishment from the bodies of decaying organisms.

Savanna — (Also spelled savannah, sabana, etc.) Grasslands (q.v.) containing numerous but isolated trees.

Scientific classification — (flora and fauna) — See *taxonomy*

Seasonality — Phenomena which show cyclic or repeated behavior according to the season.

Sedge meadow — A vegetation (usually in wet situations) consisting of low grass-like plants belonging to the family Cyperaceae, distinguished from grasses by having stems triangular in cross section.

Sediment — Any usually finely divided organic and/or mineral matter deposited by air or water in nonturbulent areas.

Seepage – The relatively slow trickling of water or other liquid from a source; a seepage lake has no visible surface inflow.

Seiche – An internal wave that oscillates in lakes, gulfs, or bays over periods of a few minutes or hours, resulting from wind, tidal forces, or (rarely) from seismic activity. Oscillation is most dramatic and most likely to cause damage after the wind has dropped.

Septic – Referring to the presence of disease-producing bacteria or other microorganisms.

Seral (stages) – Developmental temporary communities in a sere; not fixed.

Sere – A developmental series of communities which can be verified during succession (q.v.); one of a chain of seral stages containing the initial (pioneer) stage, one or more transitional stages, and a single (often hypothetical) climax stage (q.v.).

Sessile – Stationary; attached; nonmoving; in botany, non-stalked.

Seston – Particulate material including plankton, living and dead, and detritus or tripton (q.v.), retained by fine-meshed nets; a collective term designating everything that floats or is suspended in the water.

Shell fish – Aquatic animals, usually molluscs (q.v.), having an external shell or exoskeleton.

Shoal – A shallow place in a body of water; also (from "school") a mass of plankton or fish.

Shrub – A woody perennial of smaller height than a tree.

Siltation – Referring to the deposition of silt-sized (smaller than sand-sized) particles.

Silurian – The period of the Paleozoic era characterized by the appearance of land plants; also, the rocks of that period.

Slough – A wet place of deep mud or mire, or temporary or permanent lake; ordinarily found on or at the edge of the flood plain of a river.

Sludge – (Biological) The organic or mixed organic and inorganic deposit accumulating on the bottom of a stream; particle size is that of silt or clay (not sand).

Slush – Partially melted snow or ice.

Soil aggregation – The lumping together of soil particles into a coherent mass.

Soil organism – An organism ordinarily found living and reproducing in the soil.

Soil profile – The physical and chemical features of the soil imagined or seen in vertical section from its surface to the point at which the characteristics of the parent rock are not modified by surface weathering or soil processes.

Solum – Upper weathered part of the soil (A and B horizons).

Spawning beds – Those places in which the eggs of aquatic animals lodge or are placed during or after fertilization.

Species – The smallest natural population regarded as sufficiently different from all other populations to deserve a name, and assumed or proved to remain different despite interbreeding with related species.

Species – dominant – See *dominant*.

Species composition – Referring to the kinds and numbers of species occupying an area.

Species diversity – Refers to the number of species or other kinds in an area, and, for purposes of quantification, to their relative abundance as well.

Species diversity index – Any of several mathematical indices which express in one term the number of kinds of species and the relative numbers of each in an area.

Species, rare – Unusual species in an area.

Species, relict – See *relict*.

Spore – A nonsexual reproductive cell in plants.

Stability (ecological) – The tendency of systems, especially ecosystems, to persist, relatively unchanged, through time; also persistence of a component of a system; the inverse of its turnover time.

Stand – An aggregation of plants, ordinarily trees, standing in a definite limited area.

Standing crop – The biological mass (biomass) of certain or all living organisms of an area or volume at some specific time, i.e., what could be harvested.

Stenohaline – Pertaininig to organisms which can endure only a narrow range of salt in solution. Stenohaline marine organisms cannot withstand significant departures from full marine salinity, 30 to 35 parts per thousand.

Stenothermal – Pertaining to species restricted to a narrow range of temperatures.

Stratification – The natural division of a plant community into superposed strata or layers; also, division of a water body into two or more depth zones, as in "thermal" or "density stratification."

Stratum, strata – Layers, as of sedimentary or otherwise bedded rocks.

Stratification, thermal – The division of water

or air into layers (depth zones) of different temperatures and/or densities.

Stress — The result or consequent state of a physical or chemical, or social, stimulus on an organism or system; properly, a state of strain, resulting from stress; a stimulus, but medical ecology uses "stressor."

Sterilization — The killing of all organisms in an area or volume; also, the removal of the ability to reproduce.

Subclimax — A stage in a community's development, i.e., succession (q.v.) before its final (climax) stage; a community simulating climax because of its further development being inhibited by some disturbing factor (e.g., fire, poor soil).

Sublittoral — Below the lake or seashore; pertaining to the area between the low-tide mark and (say) 20 fathoms.

Substrate — The layer on which organisms grow, often used synonymously with surface of ground; also, the substance, usually a protein, attached by an enzyme; often but improperly used as a variant of substratum.

Succession — The replacement of one community by another; the definition includes the (controversial or hypothetical) possibility of "retrograde" succession.

Succession, plant — The replacement of one kind of plant assemblage by another through time.

Succession, primary — Refers to succession which begins on bare, unmodified substrata.

Succession, secondary — Refers to succession which occurs on formerly vegetated areas (i.e., having an already developed soil) after disturbance or clearing.

Suspended solids — Refers to solid (particulate) materials held in suspension; i.e., in more or less turbulent air or water, and capable of settling out when turbulence ceases.

Swamp — A flat, wet area usually or periodically covered by standing water and supporting a growth of trees, shrubs, and grasses; in contrast to a bog (q.v.), the organic soil is thin and readily permeated by roots and nutrients.

Symbiosis — The living together of dissimilar organisms by definition when the relationship is both mutually beneficial and essential.

Systems ecology — That branch of ecology which incorporates the viewpoints and techniques of systems analysis and engineering, especially those techniques having to do with the simulation of systems using computers and mathematical models.

System stability — The degree to which a system continues to function relatively unchanged when stressed (perturbed).

Synergism — The nonadditive effect of two or more substances or organisms acting together. Examples include synthesis of lachrymotors from other hydrocarbons in sunlit smog and dependence of termites on intestinal protozoans for digestion of cellulose (wood).

Taxon — Any taxonomic unit, from biotype or ecotype to phylum or kingdom.

Taxonomy — The study of principles and practice for the orderly classification of organisms.

Teleost — Pertaining to ordinary ("bony") fishes, exclusive of sharks, lampreys, gar, sturgeons, and a few others.

Terrain — A tract of land; also (terrane), its physical features with special emphasis on bedrock geology.

Terrestrial — Pertaining to land, the continents, and/or dry ground; contrasted to aquatic.

Territoriality — Any active behavioral mechanism that spaces organisms or groups apart from one another (usually shown by vertebrate (q.v.) animals).

Territory — That area which an animal actively defends. Home range (q.v.) is not necessarily territory.

Thermocline — A narrow (horizontal) zone of water in lakes and oceans with a steep temperature gradient, separating a warmer surface layer (epilimnion, epithalassa) from a cooler bottom layer (hypolimnion, hypothalassa); as a thermocline is a plane, but a zone is observed, the preferred or usual term is metalimnion (q.v.)

Thermal pollution — The excessive raising or lowering of water temperatures above or below normal seasonal ranges in streams, lakes, estuaries, or oceans as the result of discharge of hot or cold effluents into such waters.

Thermal stratification — The seasonal formation of horizontal layers of water in lakes and oceans (warm surface, cool bottom) of markedly varying temperatures, separated by a zone with a steep temperature gradient.

Thermocouple — A device used to measure temperature differences.

Thermotaxis — Directional movement induced by heat; moving toward or away from a heat source.

Thigmotropic — A response of an organism to touch, i.e., to mechanical stimulation.

Tidal marsh — Marsh land periodically inun

dated by tidal oceanic or estuarine water (i.e., salt marsh).

Tolerance — An organism's capacity to endure or adapt to (usually temporary) unfavorable environmental factors. See *ecological amplitude.*

Tolerant organism — An organism exhibiting a capacity to survive relatively large environmental changes.

Topography — Description or representation of natural or artificial features of the landscape; the description of any surface, but usually the earth's.

Toxic — Poisonous.

Trace elements — Chemical elements appearing in minute quantities in natural systems or media; may occasionally be concentrated by specific organisms. Nutrients such as phosphorus though in minute quantities, are not usually called trace elements.

Transect — A line (or belt) through a community on which are indicated the important characteristics of the individuals of the species observed; sampling along a transect may be plotless or refer to specific plots.

Transpiration — The loss of water from plants, normally as vapor.

Tripton — The nonliving component of the seston (q.v.); suspended nonliving matter in a body of water.

Trophic — Pertaining to nourishment or feeding. See eutrophic, biotic pyramid (to be carefully distinguished from tropic, responding or inclining, and topic, referring to place.)

Trophic level — All organisms which secure their food at a common step away from the first level, e.g., (1) plants, (2) herbivores, (3) carnivores.

Turbidity, — Condition of water resulting from suspended matter; water is turbid when its load of suspended material is conspicuous.

Turnover (overturn) — (limnology) Mixing of a water body from top to bottom ordinarily in fall and spring, resulting from wind action on uniformly heated or at least uniformly dense water; separates periods of stratification and may result in upwelling of nutrient-rich bottom waters. Also (systems ecology), the reciprocal of residence time, an aspect of the stability or persistence of a component (such as a species population).

Ubiquitous — Being found in many widely divergent places; able to thrive under different conditions.

Understory — Vegetation zone lying between the forest canopy (overstory) layer and the vegetation covering the ground (ground cover).

Upland — All types of land forms other than depressions (occupied by lakes, swamps) or those areas in close proximity to rivers, streams, or seas (flood plains, beaches, mud- or tide-flats, salt marshes).

Vapor — A substance in a gaseous state, i.e., neither liquid nor solid.

Vascular — Pertaining to vessels or channels for conveying fluids (as blood or sap); also, tissues supplied with such channels.

Vector — An organism that carries a disease, parasite, or infection; also (physics), a force that has both magnitude and directionality.

Vector control — Process of controlling a disease, parasite, or infection by control of the carrier.

Vegetation — Plants in general, or the total assemblage of plants, and their gross appearance as determined by the largest and most common. Flora (q.v.) is used for the list of kinds of plants.

Vegetation type — A plant community of any size, rank, or state of development.

Vernal — Of spring.

Vertebrate — Those animals possessing a spinal column or backbone, i.e., fishes, birds, amphibians, mammals, and reptiles.

Volatiles — Materials that pass into a gaseous state at ordinary temperatures and pressures; in geochemistry, substances that readily move or have moved through the earth's atmosphere.

Warmwater fisheries — The organized, sustained exploitation of populations of fishes inhabitating warm (or tropical) waters; usually implies bass, pike, etc., in contrast to salmonid fisheries.

Wastewater — Water derived from a municipal or industrial waste treatment plant.

Waste — Refuse from places of human or animal habitation; a solid, liquid, or gaseous by-product derived from human activities.

Water pollution — See *pollution.*

Water table — The upper limit of that part of the ground which is saturated with water.

Watershed — An entire drainage basin including all living and nonliving components of the system.

Wetlands — Land containing high quantities of soil moisture, i.e., where the water table (q.v.) is at or near the surface for most of the year.

Wilderness — A tract or region of land uncultivated and uninhabited by human beings, or unoccupied by human settlements.

Wildlife — Undomesticated animals; often hunted or at least noticed by man, and therefore consisting mainly of mammals, birds, and a few lower vertebrates and insects.

Wildlife enhancement — Manipulation of wildlife regions to promote increases in the amount or quality of living animals.

Wildlife habitat — Suitable upland or wetland areas promoting survival of wildlife.

Winterkill — Wildlife or vegetation dying from exposure to cold winter weather, or fishes dying from suffocation under snow-covered ice.

Woodland — Areas dominated by small scattered trees with little overlap of canopy branches, or, loosely, a small tract of closed forest.

Xeric — Characterized by or pertaining to conditions of scanty moisture supply.

Xerophyte — A plant which can subsist with a small amount of moisture, as a desert plant.

Zonation — Distinct, conspicuous layers or belts, e.g., in soils, vegetation, bodies of water, and on mountains.

Zooplankton — Small aquatic animals; see *plankton*.

Index

INDEX

A

figures, 137—138
tables, 135—136, 139
Alligator weed flea beetle, alligator weed control by, 9, 64,
    99—103, 115—122
    distribution in Southeastern United States, 111
        table, 112—113
    environmental effects of, 194
        figures, 103—106
    host specificity for, 109—110
        table, 111
    tables, 107, 111—113, 120—122
Alligator weed moth, control by, 115—122
    tables, 120—122
Alligator weed thrips, control by, 115—118
Alternanthera sp., see Alligator weed
Alternaria sp., control of water hyacinth with, 183
    table, 188
Amaranthus sp., 109—111
Amine salt of 2,4-D, see DMA
Amur, white, see White amur
Amynothrips andersoni, see Alligator weed thrips
Anopheles sp., see Mosquitos
AQI, see Agricultural Quarantine Inspection Division
Aquatic ecosystem, see Ecosystem, aquatic
Aquatic fauna and flora, effect on of dams and reservoirs,
        see also Aquatic plants, 26, 27, 33, 38—39
Aquatic plants, see also specific plants by name
    aquatic ecosystem and, 3, 7—9, 11—12
    biological control, see Biological control systems
    biological introduction into environment, 191—192
    chemical control, see Chemical control systems
    control, see control system
    distribution, tables, 53, 64, 74
    economic advantages of control, 84—85
    emersed and marginal
        India, 74
        Lower Mekong River, 53—55
        Philippines, 63—64
        United States, 165
    floating
        India, 74
        Lower Mekong River, 51, 53—54
        Philippines, 63, 64
    growth in dams or reservoirs, 38—40
        India, 69—78
        Lower Mekong River Basin, 51—60
        Philippines, 61—68
    harvesting, 141—147
    integrated control, see Integrated control systems
    marginal, see Aquatic plants, emersed and marginal
    mechanical control, see Mechanical control systems
    problems of dense infestation, 3—4, 11—12, 38—39,
        83—85, 129—130, 153, 191—192
    rooted, mechanical control, 146
    submersed
        India, 73—74, 78
        Lower Mekong River, 52—55, 57, 59
        Philippines, 63, 64
Archaeological losses, see Cultural resources
ARS, see Agricultural Research Service
Arsenic compounds, 84
Arthropod disease vectors, increased by reservoirs, 28

Arzama sp., aquatic plant control with, 56, 176, 195
Aspergillus sp., control of water hyacinth with, 183—184
    table, 188
Atomic Energy Commission, 18—19
Atriplex hastata L., host for alligator weed flea beetle,
        109—110
    table, 111
Atriplex sp., 109—111
Autecology, defined, 3, 5
Aymphaea sp., control of, 76
Azolla sp., see Water velvet

B

Bacteria, see Microorganisms
Bagoine weevil, see *Neochetina* sp.
Bamboo, 54
Bambusa sp., see Bamboo
Bank weeds, see Aquatic plants, emersed and marginal
Basin management, see River Basin Studies; River basin
        survey and assessment
Bass, large-mouth, 2,4-D residues in, 153
Bean hay, 2,4-D residues in, 167
    table, 167
Beans, snap, 2,4-D residues in, 166
    table, 166
Beauvaria sp., enemy of *Neochetina* sp., 178—179
BEE, metabolism in fish, see also PGBEE, 169—171
    tables, 170—172
Beetles, enemies of *Neochetina* sp., 178—179
Bicol Peninsula river basin development (Philippines), 61
Bilhaeziasis, 36—37
Bioassay methods, herbicides, 12
Biocides, 11—12
Biological control systems, 8—9, 39, 191—195
    alligator weed, see Alligator weed, biological control
    conflicts of interest, 192—193
    Division of Fisheries Research, 11—12
    India, 70, 75—78
    Lower Mekong River Basin, 56—59
    Philippines, 61, 64—67
    water hyacinth, see Water hyacinth, biological control
Biological oxygen demand, dams and reservoirs, 41—42
Biological resources lost by hydroelectric power projects,
        28—29
Black fly, diseases caused by, 37
Bladderwort, 53, 54, 77—78
Blight of water hyacinth, see Fungus, control of water
        hyacinth by
Bluegill, 2,4-D residues, 153, 155, 169, 171
    table, 172
    TL_m, 169
Blyxa, 53, 54
BOD, see Biological oxygen demand
Branch of River Basin Studies, see Division of River Basin
        Studies
Bucholzia philoxeroides, see Alligator weed
Bulrush, 53, 54
Butoxy-ethanol ester of 2,4-D, see BEE
By-pass channels, 40

C

D

Light and sound guides, 40
Limnophila, 53, 54
Lotus, water, see Water lotus
Louisiana program, 2, 4-D residue studies, 157
  tables, 158—159, 163
Louvre systems, 40
Lower Mekong River Basin project, 51—60
  figures, 58, 60
  tables, 53, 57
Ludwigia sp., see Water primrose
Luzon (Philippines), see Central Luzon river basin
    development (Philippines)

M

Macrophytes, 59, 65, 78, 137
Malaria, 37, 131
Manatee, water hyacinth control by, 195
Marginal aquatic plants, see Aquatic plants, emersed and
    marginal
Marsilea sp., control of, 77
Matrix analysis, 2, 4-D residue studies (Florida and
    Louisiana), 160
  table, 163
MCPA, water hyacinth control with, table, 151
Mechanical control systems
  alligator weed, see Alligator weed, mechanical control
      methods
  aquatic plants, general system, see Harvesting of aquatic
      plants
  disposal systems, 143
  equipment, 143—145
  expense of, 39
  history of, 83—85
  India, 69, 74, 76
  Lower Mekong River Basin, 51, 55
  Philippines, 61, 64—66
  water hyacinth, see Water hyacinth, mechanical control
Mekong River, see Lower Mekong River Basin project
Metabolism of 2, 4-D in fish, 169—171
  tables, 170—172
Metynnis roosevelt Eig., see Silver dollar
Microorganisms
  biodegredation by, 151, 182
  excessive growth of, 26, 27, 33, 39
Micropterus sp., see Bass, large-mouth
Milfoil, see Eurasian water milfoil
Minnow, fathead, 2, 4-D residues in, 169
Mite, see also *Orthogalumna* sp., 178—179
Morninglory, water, see Water morninglory
Mosquitos
  breeding grounds, 11—12, 28, 34, 153
  diseases caused by, 28, 37—38, 129, 131—132
Moths, see Alligator weed moths; *Arzama* sp.
Mud flats, caused by hydropower plants, 25
Multiple regression analysis of 2, 4-D residues, 171
  table, 172
Myolossoma argenteum E. Ahl, aquatic plant control by,
    9, 56—59
Myriophyllum, sp., see Eurasian water milfoil

Najas indica L., control of, 77—78
Najas sp., see Grassy naiad
Nasturtium sp., 109—111
National Environmental Policy Act of 1969, 13—16,
    29—30
National Marine Fisheries Service, 19
National Surveys of Fishing and Hunting, 20
National Water Assessment (1968), 20
National Wildlife Refuges, 17—18
  table, 18
Nechamandra sp., control of, 77
Negros Island river basin development (Philippines),
    61—62
Nelumbo sp., see Water lotus
Neochetina sp., water hyacinth control with, 56,
    173—182, 187, 194—195
  figures, 174—175
  host specificity of water hyacinth, studies of, 177—182
    tests, 179—182
  life cycle, 177—178
  natural enemies and competitors, 178—179
  table, 175
  taxonomy, 177
NEPA, see National Environmental Policy Act of 1969
Nile tilapia, see *Tilapia* sp.
Nitela sp., control of, 76
Nitrate, 34
Nitrates and nitrites, effect of on water hyacinth growth,
    136, 138
Noctuid moth, see *Arzama* sp.
Nuphar sp., specificity tests, 179, 180
Nymphaea sp., see Water lily
Nymphoides sp., control of, 76, 77

O

Office of River Basin Studies, see Division of River Basin
    Studies
Onchoceriosis, 37
Orabitid mite, see *Orthogalumna* sp.
Organic acid inhibition by 2, 4-D, 87
  table, 90
Orthogalumna sp., control of water hyacinth with,
    173—174, 176, 187
Oryza sp., 109—111
Ottelia sp., see Water lettuce
7-Oxabicyclo-(2.2.1)heptane-2,3-dicarboxylic acid, see
    Endothall
Oxygen
  deoxygenation, see Deoxygenation of water
  depletion in dams and reservoirs, see Deoxygenation of
      water
  dissolved, requirement in dams and reservoirs, 41—42,
      131, 132

P

Paired plant tests, *Neochetina* sp., 180
Pa Mong Reservoir (Thailand) project, 51—60
Pampanga River, see Upper Pampanga River basin development (Philippines)
Paper pulp from water hyacinth, 147
Paspalum, water, see Water paspalum
People, see Population
Peppers, sweet, 2, 4-D residues in, 166
  table, 166
Pests, introduction into environment, 191—195
PGBEE, aquatic plant treatment with, see also BEE, 99, 101—103
  tables, 101—102
pH, water, importance of, 27, 135, 136, 138, 171
  table, 172
Phenoxy herbicides, aquatic plant control with, see also 2, 4-D; Herbicides, 84
  alligator weed control with, see Alligator weed, 2, 4-D, control by
  organic acid inhibition by, 87
    table, 90
  water hyacinth control with, see Water hyacinth, 2, 4-D, control by
Philippines, river basin projects, 61—68
  figures, 62, 65—67
  tables, 64, 68
Phosphate, 34
Photography, aerial, in aquatic plant control, 29
Photosynthesis, phytoplankton and, 52—53
Phragmites sp., see Reed, giant
Phytoplankton, excessive growth of, 33, 52—53
Phytotoxicity of 2, 4-D derivatives, 84
Pickerelweed, 194--195
Pimephales sp., see Minnow, fathead
Pistia sp., see Water lettuce
Pithomyces sp., control of water hyacinth with, 183—184
  table, 188
Pithophora sp., control of, 76
Pituitary glands, fish, used to induce spawning in white amur, 67—68, 72, 76
Plankton, growth of, 27, 33—34, 59, 131
Planktonic algae, see Phytoplankton
Plantation Golf Course 2, 4-D residue study (Florida), 153—155
  table, 154
Plant group tests, *Neochetina* sp., 180—182
Pollution, see Air pollution; Water, pollution
Polygonum sp., see Water smartweed
Ponds, see also Dams, Reservoirs
  2, 4-D residue studies, 153—155, 171
    tables, 154, 171
Pondweeds, 64, 66, 76
Pontederiaceae, see *Pontedoria* sp.; Water hyacinth, 56
Pontedaria sp., see also water hyacinth
  disease of, 183
  host specificity of, 56, 173—176, 179—182
Pontius sp., aquatic plant control with, 59
Population, resettlement of caused by dam building, 34—36
  health hazards, 34—36

Potamogeton sp., see Pondweeds
Potato, 2, 4-D residues in, 167
  table, 166
Prairie Creek (Florida), 2, 4-D residue studies, 157—160
  tables, 162—163
Prickly pear cactus, environmental effect of, 191
Primrose, water, see Water primrose
Propylene glycol butyl ether esters, see PGBEE
Prosser (Washington), 2, 4-D residue study, 166—167
  tables, 166—167
Puccinia eichorniae, see Rust
Punta Gorda (Florida), 2, 4-D residue studies, 157—160
  tables, 160—163

R

Radioactive herbicide residues
  alligator weed, 91
  water hyacinth, 169—170
Rainbow trout, see Trout, rainbow
Redear sunfish, see Sunfish, redear
Reed, giant, 53, 54
Regional River Basin Commissions, 20
Reproduction
  alligator weed, 94—96
  water hyacinth, 132
Reservoirs, see also Dams; Hydroelectric power projects
  aquatic plants in, 38—39
  cultural losses, see Cultural resources lost by dam and reservoir building
  diseases in, 28
  eutrophication, 42—43
  geological resources and, 24—25
  fisheries, need for, 39—40
  health hazards caused by, 36—38
  India, 69—78
  land resources and, 25—26
  list of, surface area of 1000 km² or more, table 45—46
  Lower Mekong River Basin, 51—60
  Philippines, 61—68
  population resettlement caused by building, 34—36
  sedimentation in, 25—26, 39—43
  seismic effects, 44
  water conditions in, 27—28, 39—43
  water temperature in, see Water, temperature in reservoirs
Residue studies (2, 4-D residues)
  crops, 165—167
    tables, 166—167
  fish, 153—155, 169—171
    tables, 154, 170—172
  multiple regression analysis, 171
    table, 172
  slow-moving water, 157—163
    tables, 158—163
  small ponds, 153—155
    table, 154
Resettlement of people, see Population, resettlement of
Reussia rotundiflora, host to weevils, 173—174
Rheum sp., 109—111

CRC PUBLICATIONS OF RELATED INTEREST

CRC HANDBOOKS:

CRC HANDBOOK OF ENVIRONMENTAL CONTROL

Edited by **Richard G. Bond, M.S., M.P.H.**, University of Minnesota, **Conrad P. Straub, Ph.D.**, University of Minnesota, and **Richard Prober, Ph.D.**, Case Western Reserve University.

This five-volume Series is designed to deal with the major questions and problems in water and air pollution, solid waste treatment, and hospital and health-care facilities. An additional volume, the Series Index, provides immediate access to more than 12,000 entries.

CRC HANDBOOK OF TABLES FOR APPLIED ENGINEERING SCIENCE, 2nd Edition

Edited by **Ray E. Bolz, D.Eng.**, Case Western Reserve University, and **George L. Tuve, Sc.D.**, formerly Professor of Engineering at Case Institute of Technology.

This Handbook is designed to serve the practicing engineer and the engineering student with a wide spectrum of data covering many fields of modern engineering, with references to more complete sources.

CRC HANDBOOK SERIES IN MARINE SCIENCE

Three volumes are now available in this interdisciplinary Series, including data most frequently used in the physical and biological aspects of oceanography and compounds from marine organisms.

CRC UNISCIENCE PUBLICATIONS:

CADMIUM IN THE ENVIRONMENT, 2nd Edition

By **Lars T. Friberg, M.D., Magnus Piscator, M.D., Gunnar Nordberg, M.D.**, and **Tord Kjellstrom, M.E., M.B.**, all with the Department of Environmental Hygiene, The Karolinska Institute, Stockholm.

This book presents well-organized information on the toxic effects of cadmium compounds in the environment, and their effects on humans.

MERCURY IN THE ENVIRONMENT

By **Lars T. Friberg, M.D.**, Karolinska Institute, Stockholm, and **Jaroslav J. Vostal, M.D., Ph.D.**, University of Rochester.

This book deals with the health aspects of contamination of the general and industrial environment with metallic mercury and different inorganic and organic mercury compounds.

MASS SPECTROMETRY OF PESTICIDES AND POLLUTANTS

By **Stephen Safe, B.Sc., M.Sc., D.Phil.**, and **Otto Hutzinger, Ing.Chem., M.S., Ph.D.**, National Research Council, Atlantic Regional Laboratory.

All published information on the mass spectrometry of pesticides and pollutants are presented, arranged according to their chemical functionality, in addition to their mass spectra, and a discussion of the uses of mass spectrometry.

CARBAMATE INSECTICIDES: CHEMISTRY, BIOCHEMISTRY, AND TOXICOLOGY

By **Ronald J. Kuhr, Ph.D.**, New York State Agricultural Experiment Station, and **H. Wyman Dorough**, University of Kentucky.

A thorough presentation, discussion, and analysis of thousands of research articles.

PERSISTENT PESTICIDES IN THE ENVIRONMENT

By **Clive E. Edwards, M.Sc., M.S., B.Sc., Ph.D.**, Rothamsted Experimental Station, Harpenden, Herts., England.

A summary of the amounts of residues currently present in the environment, how they can be minimized in the future, the dynamics and transport of these pesticides in the environment, and a discussion of the current legislation on these chemicals.

ORGANOPHOSPHORUS PESTICIDES: ORGANIC AND BIOLOGICAL CHEMISTRY
By **Morifusa Eto,** Kyushu University, Japan.
A volume of interest to the chemist and the technician in industry.

CONTROLLED RELEASE PESTICIDES FORMULATIONS
By **Nate F. Cardarelli,** University of Akron, Ohio.
A survey of the various uses of controlled release pesticides is presented, along with laboratory and field test data from around the world.

CHLORINATED INSECTICIDES, Vols. I and II
By **G. T. Brooks, Ph.D.,** University of Sussex, England.
These volumes present complete coverage of the technology and applications of chlorinated insecticides and their environmental and biological effects.

CRC CRITICAL REVIEW JOURNALS:

CRC CRITICAL REVIEWS™ IN ENVIRONMENTAL CONTROL
Edited by **Conrad P. Straub, Ph.D.,** University of Minnesota.

Please forward inquiries to CRC Press, Inc., 2255 Palm Beach Lakes Blvd., West Palm Beach, Florida 33409 U.S.A.